WEIRD CAREERS in SCIENCE

Virus Hunter

WEIRD CAREERS IN SCIENCE

Animal Therapist

Astrobiologist

Computer Game Developer

Pyrotechnician

SETI Scientist

Virus Hunter

Volcanologist

WEIRD CAREERS in SCIENCE

Virus Hunter

Rick Emmer

CHELSEA HOUSE PUBLISHERS
A Haights Cross Communications Company ®
Philadelphia

CHELSEA HOUSE PUBLISHERS

VP, New Product Development Sally Cheney
Director of Production Kim Shinners
Creative Manager Takeshi Takahashi
Manufacturing Manager Diann Grasse
Series Designer Takeshi Takahashi
Cover Designer Takeshi Takahashi

Staff for VIRUS HUNTER

Project Management Ladybug Editorial and Design
Development Editor Tara Koellhoffer
Layout Gary Koellhoffer

A Haights Cross Communications Company ®

www.chelseahouse.com

First Printing

9 8 7 6 5 4 3 2 1

Library of Congress Cataloging-in-Publication Data
Emmer, Rick.
Virus hunter / Rick Emmer.
p. cm. — (Weird careers in science)
Includes bibliographical references and index.
ISBN 0-7910-8705-0
1. Virus diseases—Epidemiology—Juvenile literature. 2. Virologists—Juvenile literature. 3. Epidemiologists—Juvenile literature. I. Title. II. Series.
RA644.V55E46 2005
614.5'8—dc22

2005012073

All links and web addresses were checked and verified to be correct at the time of publication. Because of the dynamic nature of the web, some addresses and links may have changed since publication and may no longer be valid.

TABLE OF CONTENTS

1	Virus Hunters	7
2	Virus Hunting: The Early Years	15
3	Ebola: Hottest of the Hot Viruses	33
4	HIV: Pandemic in Progress	41
5	West Nile Virus: Hitchhiking to New York City	51
6	SARS: Nightmare in China	59
7	Looking to the Future	65
	Glossary	74
	Bibliography	77
	Websites	78
	Further Reading	79
	Organizations	80
	Index	81

Chapter 1

Virus Hunters

NO ONE SAW IT COMING. No one on the staff of Seton Hospital, in the city of Cochabamba, Bolivia, suspected that the sick young woman who had just been admitted was about to turn their world topsy-turvy and put the doctors and nurses there in grave danger. The hospital staff knew that the woman had become ill while returning from an out-of-town trip. What they did not know was that she had come down with a deadly, **contagious** disease.

OUTBREAK

While at the hospital, the young woman became sicker and sicker. Based on her **symptoms**, which included a fever and achy muscles, the doctors determined that her disease was probably caused by a **virus**. But they didn't know which one, because different illnesses can have similar symptoms (like the stomachache caused by both the stomach flu and food poisoning). In this case, the doctors thought the unlucky woman probably had either influenza or Bolivian hemorrhagic fever (BHF), a rare disease known to occur in a grassland area about 250 miles (400 kilometers) northeast of Cochabamba. Both of these killer viruses cause fever and sore muscles, but only BHF can, as its name suggests, cause **hemorrhaging**, or bleeding—lots of bleeding, from the nose, mouth, eyes, internal organs, even under the skin. Because the sick woman started bleeding, the doctors thought she probably had BHF. But before they could make a **diagnosis**, their patient died, barely two weeks after being admitted to the hospital.

Over the next two weeks, four other people became sick: the woman's aunt and father, and two of the hospital's nurses. All four of these people had spent time in the young woman's hospital room. Over the next few weeks, the aunt and father (who had gone to other hospitals when they became ill) and one of the nurses who had treated the woman died. The other nurse became very sick, but she recovered.

Up to this point, five people had become sick, and four had died. Before the **outbreak** ended, one more life would be lost.

DYING IN THE LINE OF DUTY

To investigate the two deaths at Seton Hospital, the staff decided to perform **autopsies**. Donato Aguilar, a **pathologist** (a scientist who studies diseases), was in charge of the autopsy of the nurse. With the help of another physician, he started the procedure. Part way through the autopsy, Aguilar poked his finger on the bloody tip of a scalpel. That turned out to be a deadly mistake. Within a few days, Aguilar became sick. He died two weeks later.

A virus hunter had died in the line of duty. But his death was not in vain. Authorities from the United States had sent a doctor, an expert on **infectious** diseases, to help the staff at Seton Hospital. Dr. Aguilar died shortly after the American physician, Dr. C. J. Peters, arrived. Since Seton Hospital did not have the equipment to test for BHF, Peters arranged to perform an autopsy to collect fresh samples from Aguilar's own diseased organs. These specimens were immediately frozen and shipped to a medical laboratory at the U.S. Army Medical Research Institute of Infectious Diseases (USAMRIID) at Fort Detrick, Maryland. Tests done on the specimens at USAMRIID showed that the virus that caused Aguilar's death was the Machupo virus, named after the Machupo River, which flows through an area in Bolivia where BHF appears from time to time. The Seton Hospital doctors' initial hunch had been correct. The young sick woman, as well as the other four victims, had died from BHF.

WHAT ARE VIRUSES?

Just what *is* a virus? How can something so deadly be so small (500 viruses can fit on the point of a pin!) that it can't even be seen without an **electron microscope**? (See the box on page 12.) In fact, viruses are so simple in design, some scientists do not consider them to be alive. A single virus is not even called an "organism;" it's called a "particle." A virus particle consists only of **nucleic acid** surrounded by a protective envelope, or capsule, of **protein** (Figure 1.1). Some viruses have an extra fatty envelope around the pro-

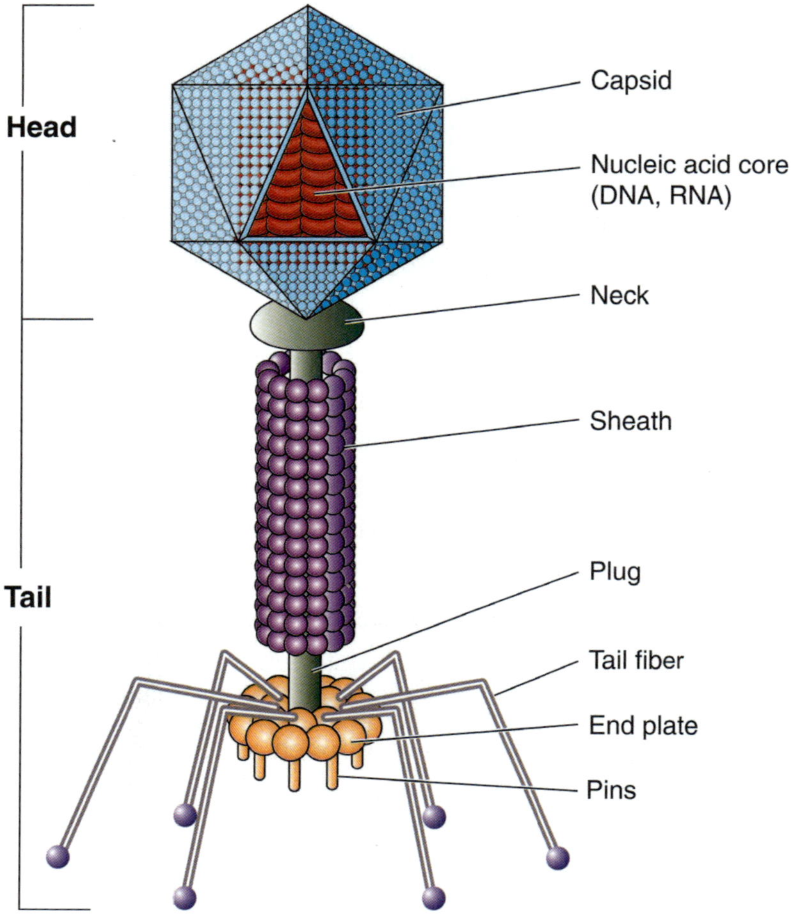

Figure 1.1 A virus is a very simple structure with just a few basic parts, as seen here. Most scientists do not even consider viruses living things, because they can't survive or reproduce on their own.

tein capsule. That's all. Viruses have no **cell membrane**, no **cytoplasm**, and no **organelles** (structures within the cell, like the **nucleus** and **mitochondria**).

The protein-enclosed bit of viral nucleic acid contains in its genes all the instructions that the virus needs to reproduce. But it doesn't reproduce on its own. In fact, it can't do anything on its own, since it doesn't have the machinery (organelles) or the raw materials (molecules like sugars, fatty acids, and amino acids) to do things that living cells do.

Then how does a virus reproduce? It sneaks into a living cell and takes control of the cell's machinery, forcing the cell to "manufacture" many copies (sometimes hundreds or thousands of copies) of its own nucleic acid and proteins, rather than performing its usual function. These new nucleic acids and proteins come together to form new virus particles, which then leave the cell, a process that often kills the cell.

Most viruses don't attack just any cell. They are usually very particular about the types of cell they invade. The cells that a certain virus attacks can usually be determined by the symptoms that the disease produces. For example, one of the symptoms of BHF is **jaundice**, a yellowing of the skin. This occurs because the virus attacks liver cells, whose job is to remove waste products from the blood. One of these waste products is a yellowish substance called bile. If bile builds up in the blood, the skin takes on a yellow tint.

VIRUS IDENTIFICATION

Modern virus hunting is a complicated process. It can involve a variety of scientists and technicians trained in many different specialties. When there is a disease outbreak, like the one in Cochabamba, the first step is to identify which virus is causing the illness. This phase of virus hunting is often performed by medical personnel trained in pathology, like Dr. Aguilar. (If the disease outbreak involves other animal species, veterinarians may be involved. If the outbreak is a plant disease threatening, say, a country's wheat crop, a **botanist** or plant pathologist will lead the hunt.)

Once an infected person (or animal or plant) is found, blood or other samples (such as saliva or urine) are collected and sent to a laboratory for study. Most of the lab equipment used to identify viruses (for example, the electron microscope) is very sophisticated and is operated by highly trained technicians. If the infected individual has died, an autopsy may be performed. (An autopsy performed on an animal is called a **necropsy**.) Once the virus causing an outbreak is identified, a **vaccine**, if available, can be given to people living in the outbreak area. This may help stop the

Tools of the Trade: Electron Microscope

Developed by the German scientists Ernst Ruska and Max Knoll in 1931, and first used to study viruses in 1933, the **transmission electron microscope** (TEM) can produce much higher magnification than an optical, or light, microscope. Whereas an optical microscope can magnify up to 2,000 times, a TEM can magnify an object 1,000,000 times. A TEM works by passing an electron beam, focused by magnets, through an extremely thin section of the specimen being viewed. The electrons that pass through the specimen form an image on a special screen or photographic plate.

Images that look three-dimensional can be obtained with a **scanning electron microscope** (SEM), a complicated computerized electron microscope developed a few years after the TEM. An SEM can magnify up to 100,000 times as it passes a narrow beam of electrons across the surface of an entire specimen, rather than a thin section of it. The electrons bounce off the surface of the specimen, scattering in all different directions. They are detected by an electronic device, which produces an image on a computer screen.

spread of the disease and prevent future outbreaks. If there is no vaccine against the virus, **virologists** will work to create one, a process that can take years to complete.

VIRUS EPIDEMIOLOGY

Identification of a deadly virus (a "hot" virus, in the lingo of virus hunters) and the development of a vaccine against it are two parts of a three-pronged attack on the disease. The purpose of the third phase is to determine the **epidemiology** of the virus. A virus-hunting epidemiologist studies the "big picture" and tries to find answers to questions such as: Who was the first person infected? Where was the person when he or she came in contact with the virus, and what was he or she doing? Did this person live in a house with other people; if so, did any of them also get sick? Were there any sick animals hanging around this person lately? The most important person for the epidemiologist to learn about is the first person to get sick. This person is referred to as the **index case**. Often, the source of the disease can be determined by finding out where the index case went and what he or she did during the days before he or she became ill. (The young woman admitted to Seton Hospital was the index case for the BHF outbreak in Cochabamba. Unfortunately, the epidemiologists trying to retrace her out-of-town trip were unable to determine how and where the woman was infected. The source of her illness remains a mystery to this day.)

Epidemiologists sometimes perform experiments to test **hypotheses** about how outbreaks develop. Their findings can then be used to stop outbreaks that are already in progress, or at least keep them from spreading. For example, one scientist studying a previous outbreak of BHF in another town in Bolivia suspected that a certain species of mouse was infecting the people with the virus. Mice were everywhere, in and around the houses, and people were getting sick all over town. The scientist persuaded the people living on one side of town to set traps to catch all the mice. Within a few weeks, people stopped getting sick. Then, a month later, he had the people on the other side of town trap all their mice. Within a few weeks, no one else had become sick. The outbreak was over, and the scientist had found the source of the virus.

As this chapter has shown, virus hunting is a dangerous and complicated process. However, with scientists, doctors, and other specialists working together, many hot viruses have been "put on ice." The next chapter goes back in time to look at a few of the victories of virus hunters.

Chapter 2

Virus Hunting: The Early Years

IF VIROLOGISTS OF THE PAST were to put together a "Virus Hunters' Ten Most Wanted List," the four viruses presented in this chapter would surely be included: smallpox, rabies, yellow fever, and poliomyelitis. Whole villages, cities, countries, and empires have been decimated by outbreaks of these diseases. **Epidemics** have killed or permanently disabled hundreds, thousands, even millions, of people. But thanks to the heroic efforts of scientists in the 18th, 19th, and 20th centuries, the viruses responsible for all of these diseases were isolated, and vaccines against all of them were developed.

As the following brief accounts show, the first virus hunters did not need lots of fancy, high-tech equipment to do their job. For example, most of the scientists mentioned in this chapter never even saw the viruses they were studying, because the electron microscope had not yet been invented. What these virus hunters did need was an inquisitive, imaginative mind. All of the scientists described below were able to use the technology available at the time (much of it very simple by today's standards) or to invent the technology they needed to get the job done.

SMALLPOX

What do Roman Emperor Marcus Aurelius, France's King Louis XV, Prince John of England, George Washington, and Abraham Lincoln have in common? They all caught smallpox. Washington and Lincoln survived the disease, but the others all died from it. And they weren't alone. From the time of the ancient Egyptian empire 3,000 years ago until the mid-1900s, historians around the world recorded one smallpox epidemic after another. The virus traveled the globe as civilizations grew and expanded. By the time this terrifying disease was finally wiped out in the 1970s, smallpox had killed countless millions of people; in the 20th century alone, it killed 500 million.

Symptoms and Outcomes

Smallpox is extremely contagious. The first symptoms of the disease occur about two weeks after exposure to the virus and include fever, headache, and weakness. A few days later, a telltale rash appears on the skin as well as the inside of the mouth and throat (Figure 2.1). Once the rash shows up, a person becomes contagious, because the rash spots, called **lesions**, contain virus particles. As an infected

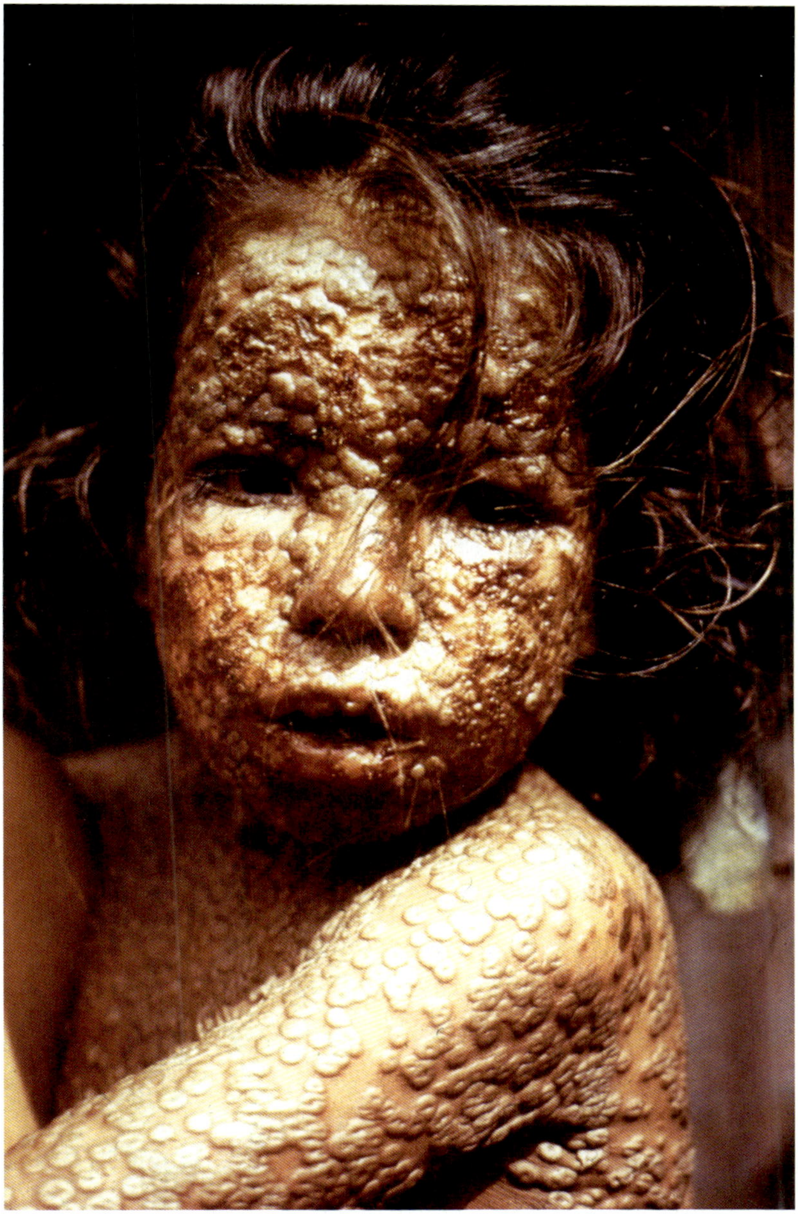

Figure 2.1 Smallpox was a horrible disease, characterized by ugly lesions on the skin, as seen on this child from Bangladesh. Victims who survived smallpox were scarred for life and often suffered permanent kidney damage and blindness.

person breathes, exhaled breath picks up virus particles from lesions in the mouth and throat and spreads them into the surrounding air. An uninfected person can pick up the virus just by inhaling some of these airborne virus particles. A person can also get the virus by touching an infected person's skin lesions, until the scabs that eventually form on them fall off. The scabs themselves are infectious, too, even *after* they fall off. No wonder this disease spread so widely!

Smallpox is deadly. Not only does it produce a rash, but it damages internal organs. It has a death rate of nearly 40%. That is, for every 100 people who become ill, about 40 will die. Those who survive will be left with ugly scars from the lesions, and many will suffer permanent kidney damage or blindness.

Thankfully, smallpox is no longer killing people by the millions. In fact, it is not killing anyone at all. The disease has been wiped out. In the 1960s and 1970s, the World Health Organization (WHO), the international public health agency of the United Nations, began a worldwide vaccination program against smallpox. It worked: The last reported case of smallpox occurred in Somalia in 1977, and the disease hasn't been heard from since. But where did the smallpox vaccine come from? To find out, we need to travel back to 18th-century England.

Edward Jenner

The year was 1796. Edward Jenner was a physician in Berkeley, located in the farmland of Gloucestershire County. England had had its share of smallpox outbreaks and epidemics over the years, so many of the townspeople of Berkeley were smallpox survivors who had smallpox scars on their skin. Curiously, the women who milked the cows at nearby farms had smooth complexions, free of pox scars. At one time or another, the dairy herds had come down with a disease called cowpox, which produces a pox-like rash on a cow's udder. Cowpox is infectious to people, so the milkmaids who milked cows that had cowpox also became infected. Fortunately, the disease is not dangerous. After a short illness that causes fever, nausea, and pox-like blisters on the hands, the milkmaids recovered with no lasting effects, except for one: They were **immune** to smallpox!

This gave Jenner an idea: Maybe he could make other people immune to smallpox by deliberately exposing them to cowpox. He decided to perform an experiment. He took some liquid from a cowpox blister on the hand of Sarah Nelmes, a local milkmaid. He then rubbed some of the liquid into a scratch on the arm of a brave young boy named James Phipps. Several weeks later, Jenner **inoculated** the boy with smallpox. Young James remained disease-free! Jenner repeatedly exposed the boy to smallpox over the next several years, and he never became infected. Edward Jenner had invented **immunization**.

In the years after Jenner's discovery, the smallpox vaccine was used in England, the United States, and elsewhere, but it took the truly amazing effort of the WHO smallpox vaccination project mentioned above to achieve the ultimate goal: wiping out smallpox worldwide.

RABIES

Rabies is a very scary disease. It could provide the perfect story line for a horror movie:

The year: 1885. The place: a farming community in the province of Smolensk, Russia, 200 miles (320 km) west of

Moscow. And . . . action! A mad, ***rabid*** *wolf, growling and snarling and foaming at the mouth, comes tearing out of the woods at the edge of town and attacks every villager it sees, biting and clawing one person after another. For two days, the villagers are terrorized by the crazed wolf until one brave man, who has already been bitten by the animal, kills it with his ax. By the time the wolf is dead, it has bitten 19 people. All of them are doomed to die slow, agonizing deaths, unless they can reach the one person in the whole world, a French scientist, who has medicine that can save them. Can they get to him in time?*

Louis Pasteur

As a matter of fact, the events described above really did occur. A rabies-infected wolf bit and infected 19 Russian peasants, and with the help of a Russian doctor, all the injured people made the 1,200-mile (1,900 km) journey to Paris, France, in the hope that the scientist they had heard about could give them life-saving medicine. The French scientist injected the peasants with a rabies vaccine that he had recently developed. All but three of the infected people survived.

The Frenchman was none other than Louis Pasteur, one of the most famous scientists of all time. He was already well known for inventing pasteurization, a process that uses heat to kill **bacteria** that spoil milk and other foods. He had also developed vaccines against deadly bacterial diseases like **anthrax**. In addition to these accomplishments, Louis Pasteur was a virus hunter.

Like Edward Jenner, Pasteur did not know what a virus was. However, he did know that some germ (microorganism), much smaller than a bacterium and invisible under the microscopes available at the time, was the cause of rabies, a terrible disease that was always fatal. Rabies didn't just kill people. Other mammals, like dogs, rabbits, and bats, could also be infected. In fact, humans usually caught rabies after being bitten by a rabid animal.

As an 8-year-old child, young Louis Pasteur had witnessed the horrible death of people who had been bitten by a rabid wolf, an experience he would never forget. To be bitten by a rabid animal was to be given a death sentence. Unless a person was lucky enough not to have received any of the rabies virus in the bite wound, he or she would become feverish and develop a nasty headache. After several days, the victim's behavior would change, as he or she became confused, anxious, and restless, and began to experience hallucinations and insomnia (an inability to sleep). A very peculiar symptom, increased **salivation**, would also appear, along with **convulsions**. As a result, the victim would produce a lot of bubbly drool (the "foaming at the mouth" common among victims of rabies). At the end, the victim would become **paralyzed** and **comatose** and then die—every single time: The death rate for rabies was 100%. As he grew up and became a scientist, Pasteur hoped that someday he would find a cure for this terrible disease.

He did. Because of the unusual behavior, **paralysis**, and coma the disease produced, Pasteur figured that the rabies germ must be infecting the brain and other parts of the nervous system. He found that he was able to infect lab animals with rabies by inoculating them with ground-up brain tissue taken from other animals that had died from rabies. He was also able to infect animals by inoculating them with saliva obtained from rabid dogs. Now that he had isolated the virus (Figure 2.2), Pasteur set to work developing a vaccine.

In earlier experiments with disease-causing bacteria,

Pasteur had discovered that the bacteria could be made weaker, to the point that they could no longer produce illness. Vaccinating a person with the weakened, or **attenuated**, bacteria would somehow strengthen the person's **immune system** so that he or she could fight off the disease if infected later on with full-strength, **virulent** bacteria. The challenge Pasteur now faced was to find a way to make a rabies vaccine by attenuating the rabies virus.

Pasteur found that he could weaken the virus by drying it out. He took a piece of the brain of a rabid rabbit and suspended it from a thread inside a sterilized bottle. In the bottom of the bottle was a substance that absorbed moisture from the air. The top of the bottle was plugged with a piece of cotton to keep out other germs and dust. As the days went by, Pasteur inoculated lab animals with samples of the brain tissue as it slowly dried out. By the 14th day, the rabies-containing brain tissue was no longer infectious.

Next, Pasteur began a series of inoculations with dogs. On the first day, the dogs were inoculated with 14-day-old rabies-infected brain tissue. On the second day, they were inoculated with 13-day-old brain tissue. On the third day, they received 12-day-old tissue, and so on, so that by the 14th day, the dogs were getting a full-strength dose of rabies virus from a rabbit that had just died. The dogs did not develop rabies. As a final test, Pasteur placed these dogs in cages with rabid dogs, which bit the inoculated dogs. None of them got ill.

Shortly after he completed his experiments, Pasteur started to use the same procedure on people who had been attacked by rabid animals. His first patient was a 9-year-old boy named Joseph Meister, who had been bitten several times by a rabid dog. The vaccination worked, and Joseph

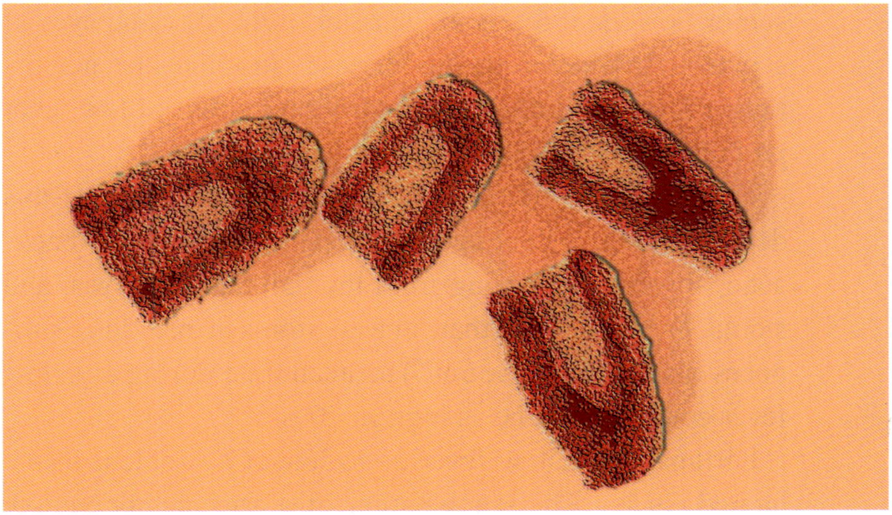

Figure 2.2 This electron micrograph shows the bullet shape of the rabies virus (magnified 90,000 times). Rabies infects the brain tissue of humans, dogs, bats, and other mammals. If untreated, the disease is always fatal.

survived. Pasteur found that as long as vaccinations were started within several days after a person was bitten by a rabid animal, he or she would not become sick. Pasteur had succeeded in making a rabies vaccine!

YELLOW FEVER

What's in a name? Sometimes a little, sometimes a lot. For example, the names "Black Death" and "plague" instantly catch our attention, since we associate these words with the horrifying epidemic that swept through Europe in the 1300s, killing more than 20 million people, one-third of the entire population of the continent. But what about "yellow fever?" It sounds pretty tame by comparison. After all, yellow is the color of the sun and daisies, and lots of diseases, including the ordinary head cold, produce a fever. But the "yellow" in the name "yellow fever" doesn't refer to the sun or flowers; it refers to the color of the skin and eyes of a person stricken with this disease. Just like the hemorrhagic fever produced by the Machupo virus, yellow fever produces jaundice, which causes liver damage.

Fever and jaundice are just two of the symptoms of this deadly disease. Other symptoms include headache; backache; shivering; bleeding from the mouth, nose, eyes, and stomach; and nausea that, due to stomach bleeding, produces black-colored vomit. The death rate during a yellow fever epidemic can be as high as 70%.

Fortunately, yellow fever is no longer a problem in the United States. The combination of a vaccine for yellow fever and the destruction of breeding grounds of the mosquito now known to spread the disease have eliminated the disease from this country. But in years past, yellow fever epidemics swept through many American cities. In the summer of 1793, Philadelphia had an epidemic that killed thousands of people and practically shut down the federal government. (Philadelphia was the nation's capital back then.) In the years before the development of a yellow fever vaccine, major epidemics hit several American cities, killing tens of thousands of people.

Yellow fever continues to be a problem in other areas of the world, especially in poor countries that cannot afford to vaccinate all their people. The WHO estimates that there are 200,000 cases of yellow fever each year in Africa and Central and South America. Of those infected, 30,000 people die—a death rate of 15%. To make matters worse, an epidemic could flare up at any time in these places.

The story of the virus hunters who set their sights on yellow fever began with the Spanish-American War in 1898. The U.S. Army was concerned that its troops in Cuba would be exposed to yellow fever, which was **endemic** to the island. As a result, the army formed the Yellow Fever Commission, a group of four physicians whose task was to study the disease and find out how it was spread. The commission, led by an army doctor named Walter Reed, went to Cuba in 1900. The Yellow Fever Commission would eventually show that a certain type of mosquito was spreading the disease. By the time the researchers' work was done, however, two of them would catch the fever, and one would die.

Walter Reed

Walter Reed was a graduate of Bellevue Medical School in New York. He became a surgeon in the army in 1875. By the time he was appointed head of the Yellow Fever Commission, he had become an expert on bacterial diseases.

The first problem to solve was how yellow fever spread. Even though the disease could spread rapidly through a crowded city, no one knew for certain if it was contagious and could pass from one person to another, as smallpox did. Research done a few years earlier by a Cuban doctor showed that the disease might be spread by a mosquito. Back in 1900, scientists did not realize that many diseases are spread from person to person by biting insects, like flies and mosquitoes. Insects that can spread disease this way are called **vectors**, and scientists at the turn of the 20th century thought it was unlikely that something so small and seemingly insignificant as a mosquito could transmit deadly diseases and cause epidemics. To test the hypothesis that yellow fever was spread by a mosquito, Walter Reed and his group performed some experiments.

First, the researchers designed a number of experiments in which several volunteers, including three of the commission members, were exposed to mosquitoes that had bitten hospital patients who were ill with yellow fever. During the course of these experiments, three volunteers came down with the disease, including two of the scientists, one of whom died; the other two volunteers survived.

Despite the loss of their partner, Reed and the other scientists continued their work. The next experiment involved two groups of volunteers. One group spent three weeks living in a mosquito-free room, wearing the dirty clothing and sleeping on the dirty beds of hospital patients who were sick with yellow fever. The other group of volunteers lived in a clean room, with clean clothing and clean beds, but they were allowed to be bitten by mosquitoes that had recently bitten patients infected with yellow fever.

The results of the experiment were very interesting. None of the volunteers from the dirty room became ill, while the volunteers who were bitten by mosquitoes in the clean room came down with yellow fever. The results were clear: Yellow fever is spread by mosquitoes and *not* from person to person through the air or by sharing clothing.

Next, the remaining members of the commission set out to determine whether yellow fever was caused by a virus. Earlier work by another scientist had indicated that the disease was caused by a type of bacterium. To see if this was true, a blood **serum** sample from a patient infected with yellow fever was carefully strained through a special type of filter with holes, or pores, so small that bacteria could not pass through. They found that the fluid that did pass through the filter was still infectious. Therefore, the disease could not be caused by bacteria; it must be caused by something even smaller—in this case, a virus.

As a result of the work of Walter Reed's Yellow Fever

Commission, people living in areas where yellow fever was endemic began to kill mosquitoes and destroy the insects' breeding sites (such as small puddles or containers filled with stagnant water). As a result, the number of mosquitoes decreased, as did the number of people who became ill with yellow fever. Because of Reed's contributions to stopping the spread of yellow fever, the Walter Reed Army Medical Center in Washington, D.C., was named in his honor.

In 1937, a scientist named Max Theiler developed a yellow fever vaccine, for which he was awarded the Nobel Prize in 1951 (Figure 2.3). Thanks to the work of Walter Reed, Max Theiler, and other virus hunters, yellow fever is not as common as it used to be. However, like rabies, it's still out there. Epidemiologists fear that sooner or later, another epidemic will occur. Only time will tell.

Figure 2.3 Max Theiler (seen here) picked up the work on yellow fever where Walter Reed left off. He won the Nobel Prize for the yellow fever vaccine he developed.

POLIOMYELITIS

Poliomyelitis is another deadly viral disease that, thanks to the work of 20th-century virus hunters, is no longer a health problem in the United States. Nicknamed "polio," this disease is caused by an infection of the nerve cells of certain parts of the brain and spinal cord that control muscle movement. As the virus kills these cells, the muscles they control may stop working, resulting in paralysis and sometimes death. Because polio much more often strikes children than adults, it is also known as infantile paralysis. Although polio has been around for thousands of years, it wasn't very common until the beginning of the 1800s, when, for some reason, outbreaks in Europe and the United States became more frequent. As time went by, more and more outbreaks turned into epidemics, and each new epidemic seemed worse than the one before.

The worst polio epidemic in the United States occurred in 1952, when 58,000 people became sick. Of these, 3,000 people died, and nearly 40% of the survivors were paralyzed. Fortunately, virus hunters had been chasing this disease for many years, and they were just about ready to deliver polio a knockout punch.

Jonas Salk

In the early 1900s, scientists in Paris succeeded in isolating the polio virus. Developing a good vaccine against polio proved to be difficult. It took scientists a while to discover that there are actually three different types, or strains, of the polio virus. Once they figured out that a vaccine developed for just one of the strains would not protect people against the other two, researchers set to work to create a vaccine that would protect against all three strains.

The first scientist to develop a successful polio vaccine was Jonas Salk, a physician at the University of Pittsburgh. Salk figured out how to make a vaccine using a technique called **tissue culture**, in which cells of organs of the body (like the liver or kidneys) are artificially grown on glass plates (see the box below). Salk was able to grow the polio virus in tissue cultures made from monkey kidneys. The virus obtained from these cultures was then killed in a solution of a poisonous chemical called formaldehyde. Using the dead virus, Salk was able to make a polio vaccine that could be injected into a person's body through a needle. This type of vaccine, which uses killed viruses, is called an **inactivated** vaccine.

In 1953, Salk's vaccine was tested on hundreds of thousands of children and found to be safe and effective. In 1955, the vaccine was injected into millions of children all

Tools of the Trade: Virus Culture

In the 1940s, a team of three scientists led by John Enders, a virologist at Harvard University, developed a technique called tissue culture, in which individual cells from different organs (such as the spleen or kidneys) could be kept alive outside of the body. When placed on a surface of glass or cellophane and soaked in a nutrient solution that contained certain vitamins, sugars, salts, and amino acids, a single cell would reproduce over and over, forming a sheet of cell tissue. If the cultured tissue was then exposed to a virus, the virus would invade some of the tissue cells, reproduce, and spread to surrounding cells, eventually producing a large number of virus particles, which could then be collected and either inactivated or attenuated to make vaccines.

over the United States (Figure 2.4). The results of this huge vaccination program were very impressive. In 1954, the year before the program started, there had been 38,000 cases of polio. Over the next several years, the number of new cases dropped to fewer than 5,800 per year.

Unfortunately, several of the children who received the vaccine during the first year of the immunization program came down with polio, and some of them died. An investigation showed that the doses of the vaccine they received were not prepared properly and contained some virus particles that had not been killed. For this and other reasons, some researchers decided to try to develop a better vaccine.

Albert Sabin

Albert Sabin, a doctor studying polio at the University of Cincinnati, wanted to attack the disease from a different direction. Unlike Salk's inactivated vaccine, which used killed viruses, Sabin's vaccine was going to use a live, but weakened, form of the virus. Like Salk, Sabin grew the polio virus in monkey kidney tissue cultures. He found that he could attenuate the virus by culturing it at temperatures cooler than normal body temperature.

Sabin thought that an easily swallowed oral vaccine would work better than Salk's vaccine, which had to be injected into the skin. Scientists had discovered earlier that the polio virus enters the body through the mouth and ends up in the digestive system, where it reproduces and then spreads to the brain or spinal cord. Sabin felt that an oral vaccine that followed the route normally taken by the virus would be a better way to vaccinate people. Using his attenuated virus, Sabin developed an oral polio vaccine that could be given to children in a small, sweet-tasting cube.

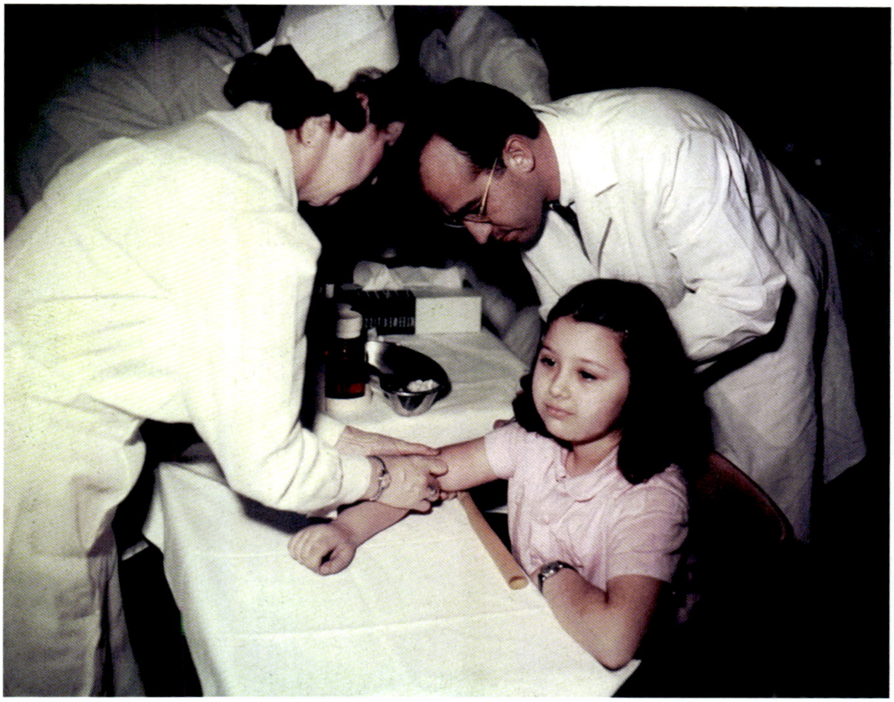

Figure 2.4 Jonas Salk developed the first successful polio vaccine. Here, a nurse is assisting him as he injects the inactivated vaccine into a child's arm.

Using the Vaccine

Once Sabin's vaccine was found to be safe and effective, the federal government had to decide which vaccine to use: Salk's or Sabin's. In 1960, the government chose to stop using the injectable Salk vaccine and start using the oral Sabin vaccine, which could be easily given to children. Every now and then, however, a child who received the Sabin vaccine came down with polio, because the virus in the vaccine was still alive. As a result, the government switched back to using the Salk vaccine. Today, some countries use Salk's vaccine while other countries use Sabin's. Worldwide, polio has become much less common than it used to be. It no longer occurs in the United States and several other countries, and it is hoped that polio will follow in the footsteps of smallpox and soon be wiped out.

As this chapter has shown, virus hunters have successfully combated a number of dangerous viral diseases over the last three centuries. But as the rest of this book will show, their work is far from done.

Chapter 3

Ebola: Hottest of the Hot Viruses

OVER THE LAST 30 YEARS, several new, dangerous viral diseases have appeared, or **emerged**, in places all over the world. They seem to arise out of thin air, infecting and sometimes killing hundreds, thousands, even millions, of people. Some of these emerging diseases are incredibly "hot": They are so infectious, contagious, and virulent that they seem to burn themselves out before they can grow into a big epidemic.

OUTBREAK IN AFRICA

Many of the hottest viruses, like the Machupo virus in Bolivia, produce hemorrhagic fever. The hottest hemorrhagic fever virus of all emerged in the 1970s in central Africa, first in the country of Sudan and several weeks later in neighboring Zaire (now known as the Democratic Republic of Congo). In both countries, deadly epidemics mysteriously flared up, quickly killed hundreds of people, and then just as mysteriously went away. As European, American, and African virus hunters investigated the epidemics and revealed just how horrifying and deadly the disease was, the virus responsible for it became known around the world. Scientists named it after an African river that flows near the sites of the two epidemics, a river named Ebola.

TERROR IN SUDAN AND ZAIRE

The epidemic in Sudan started as an outbreak in the small town of Nzara on June 27, 1976. The index case was a strong, healthy young man named Yusia, who worked at a cotton factory. Over a three-day period, Yusia became very sick. He developed a severe headache, fever, nausea, a sore throat, a very dry mouth, and sore muscles in his neck, chest, back, and legs. Yusia was so miserable that his brother brought him to the Nzara hospital. Yusia's condition quickly got worse. He developed diarrhea and began vomiting and bleeding from his mouth and nose. Finally, as author Frank Ryan wrote, "[t]he flesh shrank about his bones until his face resembled a skull-like mask, with eyes that were sunken and staring. Death came as a merciful release on July 6."

Over the next several weeks, the virus spread to nearby villages. More people became sick with the fever and went to the hospital in Nzara. Then the outbreak took off like a wildfire. More and more ill people reported to the Nzara hospital and many others went to the hospital in the nearby town of Maridi. The disease spread to the nurses and doctors. People were dying in both hospitals. So many people died at the Maridi hospital that the maternity ward was used as a morgue to hold all the dead bodies. By the time the epidemic burned itself out at the end of November, 284 people had been infected with hemorrhagic fever, and 151 had died—a death rate of 53%. But as bad as this epidemic was, the one in Zaire was worse.

The epidemic in Zaire started in late August in the town of Yambuku. The index case was a sick schoolteacher who went to get medical help at the **dispensary** at the town's Roman Catholic mission on August 26. He died there of hemorrhagic fever on September 9. The virus spread to nearby villages, and more and more sick people went to the mission for medical help. One by one, the nuns who worked there became ill, just like the doctors and nurses in Nzara and Maridi. By the time the epidemic ended in early November, 280 out of 318 infected people died—an incredible death rate of 88%!

Because the governments of Sudan and Zaire did not have the equipment and trained scientists to investigate the epidemics, other countries were asked for help. By the time authorities in Europe and the United States became aware of the problem, organized research teams, and sent them to the two stricken countries, the epidemics had already started to fizzle out. But the scientists learned as much as they could, hoping that the knowledge they gained could be used to snuff out future outbreaks. What they found out was fascinating—and scary.

STALKING EBOLA

Two of the first virus hunters to arrive in Sudan were British physicians. Cenydd Jones and David Smith visited Maridi and Nzara, collecting information about the index case (the man Yusia) and the spread of the disease into neighboring villages. Dr. Smith also collected blood samples from patients at the hospitals in Nzara and Maridi and tissue samples from the livers of some of the dead victims. The samples were carefully packaged, frozen, and shipped to a special virology laboratory in England. Some of the samples were sent to Dr. Karl Johnson at the Centers for Disease Control and Prevention (CDC), the U.S. national public health agency.

Johnson sent the samples to researcher Patricia Webb, who was able to grow the killer virus in cell cultures. She then gave a small sample of virus-infected cells to virologist Fred Murphy, chief of the branch of viral pathology at the CDC. Using an electron microscope, Dr. Murphy was able to take pictures of the virus. The skinny, wavy, wormlike virus he saw (Figure 3.1) looked exactly like only one other virus known to science: the Marburg virus.

Named after the city of Marburg, Germany, where it emerged in 1967, this virus caused a deadly outbreak of hemorrhagic fever in workers at a laboratory that produced polio vaccines. The lab housed monkeys that were kept as a source of kidney tissue used to culture the polio virus. Tests performed on the monkeys showed that they were infected with the same virus as the sick workers. The monkeys had been imported from Africa, from an area only 500 miles (805 km) from Nzara, Sudan!

The CDC scientists were almost certain they had finally identified the killer virus from Maridi and Nzara as the Marburg virus. Imagine their surprise when Patricia Webb, performing sophisticated chemical tests on the new virus, showed that it was *not* identical to the Marburg virus: It was a brand-new virus, and it was given the name Ebola.

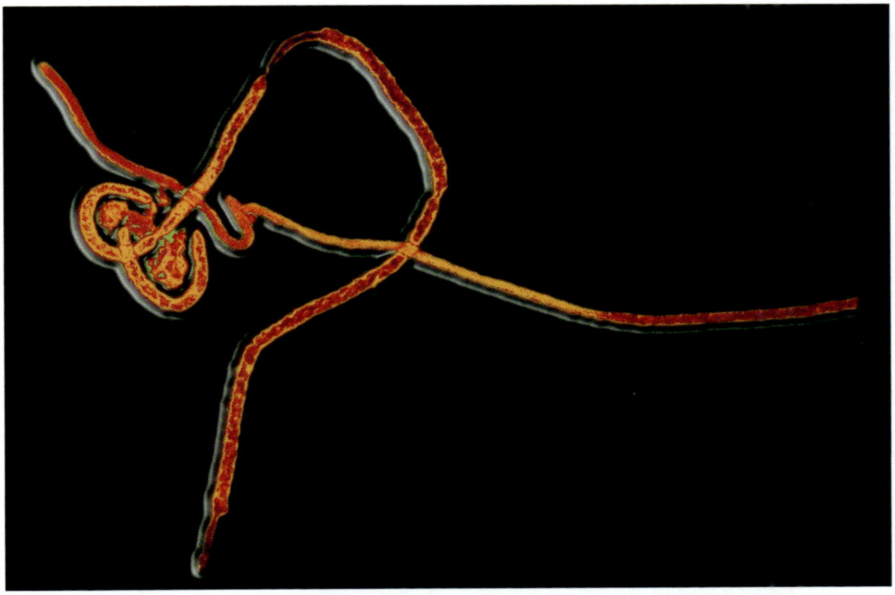

Figure 3.1 **This electron micrograph shows the worm-like appearance of the Ebola virus, magnified 65,000 times. Ebola and the nearly identical Marburg virus cause deadly hemorrhagic fever in humans and monkeys.**

Scientists now knew what they were looking for. Two teams of virus hunters were quickly sent to Africa. Studies performed by these and other groups explained why the Ebola virus had spread so rapidly. The scientists found that in both areas struck by the Ebola epidemic, people performed a funeral ceremony that involved cleaning and touching the dead body before it was buried. Anyone who touched the virus-contaminated blood of a dead person during this ritual was likely to become infected. If that person then died, anyone who touched him or her during the funeral could also become infected, and so on. Ebola was literally jumping from funeral to funeral (Figure 3.2).

Researchers also found that hospital staff in the area had so little equipment that they were reusing syringes and needles and were not always sterilizing them properly between patients. No wonder this virus had spread so fast!

One question that was never answered was: Where did the Ebola virus come from? Searching for an Ebola vector,

Figure 3.2 A victim of Ebola hemorrhagic fever is buried during a funeral service in Africa. The traditional funeral service involves washing and hugging the dead body. People who take part in this ritual may come into contact with the victim's blood, exposing themselves to the contagious, deadly virus.

virus hunters caught hundreds of animals, including some of the countless rats and bats that lived in the Nzara cotton factory. None of the animals was found to have Ebola. Even now, some scientists remain in Africa, trying to find the virus's "hideout." It's the biggest of the many mysteries that still surround the Ebola virus.

EBOLA IN THE UNITED STATES

As virulent and contagious as it is, Ebola is the last virus anyone would want to see cropping up in a large city where millions of people live and work close together. But in 1989, that's exactly what happened. And it happened right here in the United States.

Hazelton Research Products is a company that quarantines monkeys imported from other countries for use in medical research. It is located in Reston, Virginia, a suburb of Washington, D.C. When monkeys that had been received in a shipment imported from the Philippine Islands began to die from what appeared to be a form of hemorrhagic fever, Hazelton's veterinarian became suspicious. Some of the dead monkeys were carefully transported to the BSL-4 lab at nearby Fort Detrick, Maryland (see box on page 40). Tests performed on tissue samples showed that the monkeys had died from Ebola fever! Authorities had no choice but to kill and safely dispose of all the remaining monkeys, and then sterilize the entire building where the monkeys had lived.

Four of the workers at Hazelton also became infected with the virus. Fortunately, none of them became ill. Apparently, this form of Ebola was not dangerous to humans like the African form was.

Not all deadly viruses produce such obvious symptoms as those found with hemorrhagic fever. Some are downright sneaky and can infect a person for years before any clear symptoms surface. If such a virus is contagious, it can spread to a large number of people before anyone knows it's there. One such virus that sneaked into the United States in the 1980s has since spread just about everywhere, creating a worldwide epidemic, or **pandemic**. As the next chapter shows, this disease has been appropriately nicknamed "the current plague."

Tools of the Trade: Biosafety Level 4 Laboratory

Scientists studying dangerous **microbes** need to be protected against exposure. There are only a few laboratories in the world designed to be safe enough to work with the hottest viruses.

Microbiology labs can be classified into four categories: Biosafety Levels 1, 2, 3, and 4. Biosafety Level labs 1, 2, and 3 are used for working with microbes that are either totally harmless or produce diseases that are preventable or treatable with vaccines or other medicines. A Biosafety Level 4 lab (abbreviated BSL-4) is designed for working with microbes like Machupo, Marburg, and Ebola viruses, for which no vaccine or treatment is available. Here, microbe cultures are grown in special cabinets in which air can move in only one direction: from the lab room into the cabinet, and then out through a special duct. That way, any microbes that become airborne get sucked out the duct and can't float around the room.

BSL-4 lab researchers wear special protective suits that look like an astronaut's space suit, and they breathe fresh air pumped directly into the suit. When they work with their microbes, they stick their already-gloved hands into rubber gloves mounted right onto the clear front of the cabinet that holds the microbe cultures. That way, they never touch the cultures with their bare hands.

Chapter 4

HIV: Pandemic in Progress

WHAT WOULD BE THE CHARACTERISTICS of the most dangerous virus imaginable? Obviously, infection with the virus would always result in death, like an untreated rabies infection. Like Machupo and Ebola, this virus would be contagious, and there would be no cure or vaccine for it. It would not produce any obvious disease symptoms for a long time, perhaps for several years; that way, it would have time to spread to many people before anyone knew it was there. Finally, to guarantee survival, it would sneak its way into the body's immune system and destroy the system's defensive disease-fighting cells.

The scariest part of all is that this virus actually exists. It's called the human immunodeficiency virus, HIV for short. The disease it causes is called acquired immunodeficiency syndrome, or AIDS.

THE SNEAKY VIRUS

HIV is a master of disguise. This virus attacks the immune system by invading certain disease-fighting cells called T4 cells. But before the virus takes over a T4 cell's machinery to produce more virus particles, it pulls a neat disappearing act that makes it invisible to the rest of the immune system.

HIV is an unusual type of virus called a **retrovirus** (Figure 4.1). Whereas the genetic material of cells and most viruses is made of deoxyribonucleic acid (DNA), the genetic material of some viruses, including retroviruses, is made of ribonucleic acid (RNA). When a retrovirus enters a cell, it makes a DNA copy of its RNA. This DNA copy then attaches to the cell's own DNA. From there, disguised as cell DNA, it directs the production of more virus particles and eventually kills the cell.

HIV always wins its battle against the T4 cells. It may take months or years, but eventually, all the T4 cells are destroyed. Once that happens, the whole immune system is crippled. Why? T4 cells are the "lookouts" in the army of cells that make up the immune system. When a T4 cell detects a **pathogen**, it releases a chemical that alerts the rest of the immune system that an invader is present. With no T4 cells to release that signal, the immune system cannot be "turned on" to defend against pathogens.

Once HIV weakens the immune system, pathogens that a healthy immune system would normally destroy can easily invade the body, causing out-of-control **opportunistic**

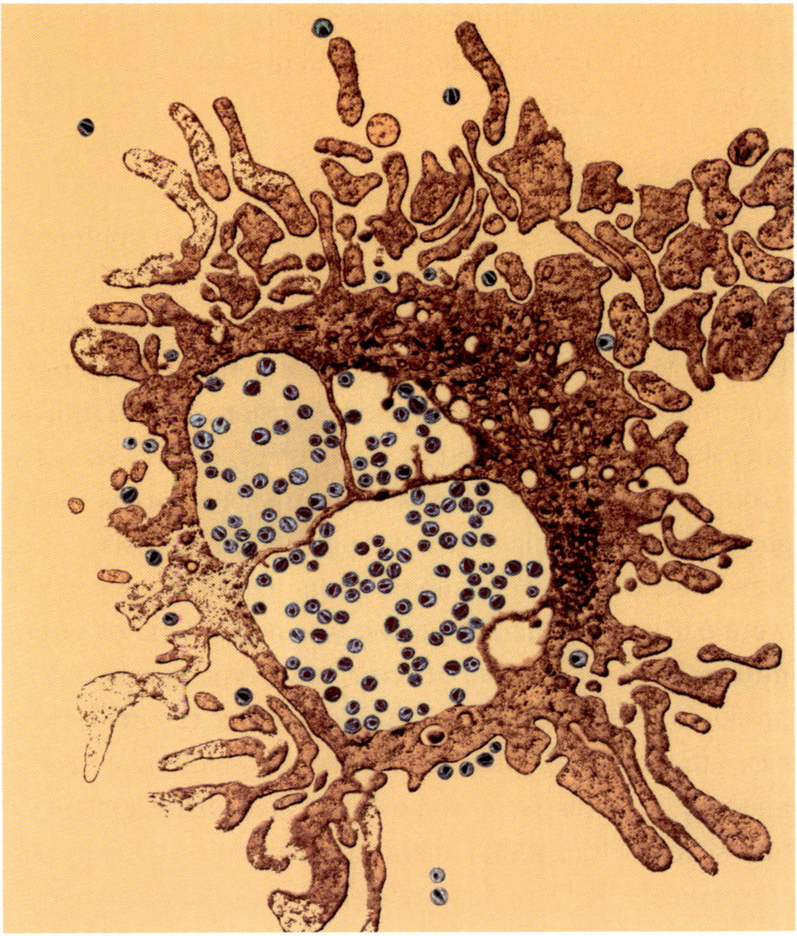

Figure 4.1 The T cell in this electron micrograph (magnified 25,000 times) is infected with numerous tiny, circular HIV retrovirus particles (seen in blue). HIV eventually kills all the body's T cells, allowing other pathogens to produce the out-of-control opportunistic infections commonly seen in AIDS patients.

infections. In fact, HIV itself does not kill anybody: It is opportunistic infections that kill people with AIDS.

Since its emergence in the early 1980s, HIV has spread to so many people in so many countries that it may end up killing more people than any other disease in history.

Approximately 38 million people worldwide were infected with HIV by the end of 2003, including nearly one million in the United States.

THE VIRUS EMERGES AND SPREADS

AIDS first appeared in the United States in the autumn of 1980 in Los Angeles, California. Over a period of several weeks, five very sick young men were admitted to the UCLA Medical Center. They had a number of serious opportunistic infections caused by normally harmless microbes, and their blood contained few, if any, T4 cells. More and more sick men with similar symptoms began showing up at hospitals in Los Angeles, San Francisco, and New York City. Two common symptoms of these patients were particularly alarming. One was a rare type of **pneumonia**. The other was an extremely rare type of **cancer** that produces lesions on the skin. As doctors reported these cases to the CDC, epidemiologists saw a pattern: The vast majority of cases were homosexual men. It looked like a new sexually transmitted disease had emerged. The disease was named GRID, or "gay-related immune deficiency."

As more cases were reported, it became clear that this mysterious ailment was not just limited to homosexual men. Many of the sick people who began to show up at hospitals, especially in New York City, were not gay; they were drug users who had apparently spread the disease by sharing drug-injecting needles. And not all of the sick were men: A number of them were women, some of whom had been pregnant and had infected their babies before, during, or after birth. As a result, scientists gave this disease a new name, the one by which we now know it: AIDS.

By 1982, AIDS had emerged in Europe and on the

Caribbean island of Haiti. By 1983, it was present in several nations of southern Africa. AIDS was no longer just an American epidemic; it was turning into a pandemic.

VIRUS HUNTERS SET THEIR SIGHTS ON HIV

Research into the cause of AIDS took a while to "get into gear," because the federal government did not want to make the unpopular decision to spend millions of dollars studying a disease that at first appeared to affect only homosexuals and drug addicts. Fortunately, the government's attitude changed as the disease began to infect people from all walks of life, including newborn babies and people who received blood **transfusions**. Two research teams, one in the United States and the other in France, became involved in a race to be the first to identify the cause of AIDS.

Robert Gallo and Luc Montagnier

Robert Gallo was the right man in the right place at the right time (Figure 4.2). In 1982, Gallo was a cancer researcher at the National Institutes of Health (NIH) in Bethesda, Maryland. The way AIDS victims' immune systems were weakened was similar to what happened in patients suffering from a cancer-causing retrovirus that Gallo had recently discovered and named human T-cell leukemia virus, or HTLV. Gallo thought there was a good chance that a similar retrovirus might be the cause of AIDS.

There is one surefire way to detect a retrovirus. A certain protein, an **enzyme** called reverse transcriptase, or RT, is produced only by retroviruses. RT is needed to help make the DNA copy of the virus's RNA. Gallo's team of virus hunters set out to grow the mystery virus in T4 cell cultures so they could test for RT. This was easier said than done,

Figure 4.2 Luc Montagnier (left) and Robert Gallo (right), leaders of the French and American research groups that identified HIV in 1984, congratulate each other after sharing the Prince of Asturia Prize for Scientific Research and Technology awarded in Spain in 2000.

since the virus kept destroying the cell cultures. Finally, the researchers were able to grow enough of the virus to test for the presence of RT.

Gallo's hunch was right: RT was found in the cultured T4 cells. Further testing showed that the virus in the cell cultures was not HTLV, the only other retrovirus known to infect humans. This was a brand-new retrovirus!

At the same time, researchers at the Pasteur Institute in France obtained similar results. A team of virus hunters led by the French scientist Luc Montagnier also found a way to culture the mystery virus in T4 cells and was able to detect the presence of RT.

The HIV retrovirus was finally identified in 1984. An assault on AIDS could now begin.

ASSAULT ON AIDS

With the discovery of the HIV retrovirus, scientists were optimistic about developing an AIDS vaccine. Some, including Robert Gallo, predicted that a vaccine would be available within just a few years. That did not happen.

Work on an AIDS vaccine

Scientists soon learned that HIV is a tricky virus to work with, mainly because it keeps changing. Mistakes frequently occur when the retrovirus makes the DNA copy of its RNA. This results in **mutations**, some of which change the appearance of the virus particle's protein envelope. A vaccine developed for the original form of the virus may not protect against a new, mutated form. Because it has such a high mutation rate, the virus is able to stay one step ahead of the scientists' vaccines. Virologists are working on ways to get around this problem. One of the most promising strategies involves designing a vaccine using bits of the DNA that the virus produces.

Treatment for AIDS Victims

Dr. Jayo Kidenya: Searching for a Cure

By 1984, AIDS was rapidly spreading in the African nation of Uganda. Dr. Jayo Kidenya was at the hospital in the village of Bukoba. He was very concerned about the AIDS epidemic: "It pains me to care for an AIDS patient. It really pains me. Because whatever I give I know it is not helping the patient. . . . I don't fear contracting the disease, but it pains me to know that whatever I do . . . it's useless."

It's easy to understand Dr. Kidenya's frustration. There was no cure for AIDS in 1984, and there is still no cure. However, once the HIV retrovirus was discovered, scientists went to work developing medicines to treat AIDS patients and to slow the spread of the disease. In fact, an American scientist would soon come to Dr. Kidenya's country to do battle against HIV.

Dr. Laura Guay: Stopping Mother-to-Child Transmission

In 1987, physician Laura Guay came to Uganda to help start research into the treatment of AIDS patients there. Guay spent three years working with scientists studying the AIDS epidemic. Guay, a **pediatrician**, was especially interested in finding a way to stop the spread of AIDS from pregnant women to their babies.

In 1996, Guay got her chance. She became involved in another research project in Uganda, using a drug called nevirapine to stop the transmission of HIV from pregnant women to their newborn babies. The results of her research were encouraging: By giving pregnant women one injection of nevirapine during childbirth, and then giving one injection of the drug to their newborn babies, the rate of transmission of the virus from mother to baby was reduced by 50%.

Drugs to Help AIDS Patients

Other scientists have been hard at work developing drugs that slow the progress of the disease in AIDS patients, so that these people can live longer, more productive lives.

One of these medicines, AZT (azidothymidine), slows down the virus by interfering with the production of retroviral DNA. Another type, protease inhibitor, interferes with the assembly of the protein envelope of new retrovirus particles. None of these medicines by itself works for long, though, because the constantly mutating virus eventually "outsmarts" it. Luckily, researchers have found that when AIDS patients are given a "cocktail" treatment of two or

Tools of the Trade: Enzyme-Linked Immunosorbent Assay

Since people infected with HIV may not show symptoms of AIDS for several years, scientists who study the disease and doctors who treat patients need a quick and reliable way to determine if a person is infected with the virus. The most commonly used procedure for detecting HIV is called enzyme-linked immunosorbent assay, or ELISA.

To perform ELISA, a laboratory technician uses a special tiny container that is coated with a thin film of a substance that contains proteins from the HIV protein envelope. The technician first rinses the container with a sample of a patient's blood. If the patient has HIV, his or her blood will contain **antibodies**, special proteins produced by the immune system, that will stick to the HIV proteins in the film. The container is then rinsed with a solution containing a "sticky" chemical that attaches only to HIV antibodies. This sticky chemical has an enzyme in it that turns a final solution, which contains a special chemical that changes color, from clear to green, when it is added to the container. Therefore, an ELISA test performed on a blood sample from a person infected with HIV will produce a green color in the final solution, whereas blood from an uninfected person will cause no color change.

more of these medicines at the same time, it takes the virus a lot longer to become resistant to the medicines.

Scientists will undoubtedly develop more medications for AIDS patients. Someday, they may even find a cure. In the meantime, however, they will have to be prepared to deal with other viruses as they emerge.

Chapter 5

West Nile Virus: Hitchhiking to New York City

THE SUMMER OF 2002 was not kind to the birds of Ohio. West Nile virus, an Old World virus that had somehow crossed the Atlantic Ocean and emerged in New York in 1999, had now reached Ohio and other Midwestern states, killing birds by the thousands. The Kenneth A. Scott Wildlife Rehabilitation Center at the Lake Erie Nature and Science Center in Bay Village, a small town west of Cleveland, was swamped with sick birds brought in by concerned citizens. Every day, people dropped off birds of all sorts—crows, blue jays, owls, hawks, ducks—suffering from symptoms characteristic of **encephalitis**, the inflammation of the brain caused by West Nile virus: The poor birds were uncoordinated and unable to fly. In fact, they could barely walk, stumbling around with droopy heads, as if they were drunk.

Megan Tadiello, Wildlife Rehabilitation Coordinator for the Rehab Center, had never seen anything like this before, as dozens and dozens of sick birds were brought in week after week: "It literally drove me and my partner to tears, every single day, because it was just overwhelming." A great horned owl and a red-tailed hawk that lived in outdoor exhibits at the Nature Center also became sick, probably having been bitten by infected mosquitoes, the vectors that spread West Nile. They, like so many of the dying wild birds brought to the Nature Center, had to be **euthanized**, because there is no cure for this deadly disease.

Even though Tadiello was suspicious that West Nile was killing the birds, she needed to know for sure. She sent tissue specimens from some of the dead birds to a laboratory equipped to test for West Nile. The test results came back a few days later, confirming Tadiello's suspicion: West Nile virus was present in all of the specimens.

The identification of West Nile virus in the birds at the Nature Center was done quickly, in a matter of a few days, because the experts knew what they were looking for. When this virus first emerged in the United States three years earlier, however, it took several weeks for scientists to identify the virus in the dead birds littering the sidewalks and streets of New York City and its suburbs. It is interesting and important to find out why it took so long. After all, West Nile virus doesn't just kill birds. It kills people, too.

AN UNUSUAL EPIDEMIC STRIKES NEW YORK CITY

Few animals are as hearty and adaptable as crows. So in early August of 1999, when wild crows started dying all over the New York City borough of Queens and at the nearby Bronx Zoo, it was obvious something was not right. Workers at the Bronx Zoo were especially concerned because of the possibility that the crows might have a contagious disease that could spread to the zoo's bird collection. Someone needed to find out what was going on.

Birds Are Dying

That "someone" turned out to be Tracey McNamara, the veterinarian in charge of the Bronx Zoo's Pathology Department. Dr. McNamara performed necropsies on some of the dead birds and looked at samples of the birds' tissues under the microscope. She found lesions in the brain tissue, which suggested that the birds had some form of encephalitis.

Within several days, the zoo's birds started to die. Having heard on a news broadcast that some New York City residents had developed symptoms of the viral disease called St. Louis encephalitis, or SLE, Dr. McNamara immediately became concerned. Could the same virus be infecting both birds and people? She decided to ask the experts in insect-transmitted diseases at a CDC laboratory in Fort Collins, Colorado. Their answer: No, SLE doesn't kill birds; the birds must have some other disease. Meanwhile, more of the zoo's birds were dying.

McNamara was not convinced by the CDC's reply. She had a hunch that there *was* a connection between the sick people and the sick birds. To prove it, she sent tissue samples from some of the dead birds to the National Veterinary Services Lab in Ames, Iowa. From those samples, technicians there were able to culture a virus that, when viewed under the electron microscope, looked very much like the SLE virus! McNamara's hunch appeared to be right.

Meanwhile, more of the zoo's birds had died; by summer's end, 24 of them would die or have to be euthanized.

People Are Dying

The first two patients with symptoms of human encephalitis (fever, confused mental state, weak muscles, and uncontrollable shaking) were admitted to Flushing Hospital Medical Center in Queens at almost the exact same time that Tracey McNamara noticed the first dead crows at the Bronx Zoo. The number of patients quickly jumped to eight.

In the meantime, in order to hunt down the type of mosquito known to be a vector for SLE, the American Museum of Natural History in New York City sent Varuni Kulasekera, an **entomologist**, to the area in Queens where several of the hospitalized patients lived. Kulasekera discovered that this type of mosquito (Figure 5.1) had been using water buckets and birdbaths in people's yards and gardens as breeding sites.

By now, lab tests had shown that the patients at Flushing Hospital probably did have SLE. Yet, something wasn't quite right. Doctors at the hospital noted that not all of the patients' symptoms were typical of people infected with SLE. They wondered if there could there be some other, rare encephalitis virus out there. They also wondered about the sick and dying birds that people were finding all around: Could there be a connection?

Birds *and* People Are Dying

By mid-September 1999, officials at both Flushing Hospital and the Bronx Zoo were concerned that there was a virus other than SLE on the loose, infecting both people and birds. Up until now, the CDC had stuck by its statement

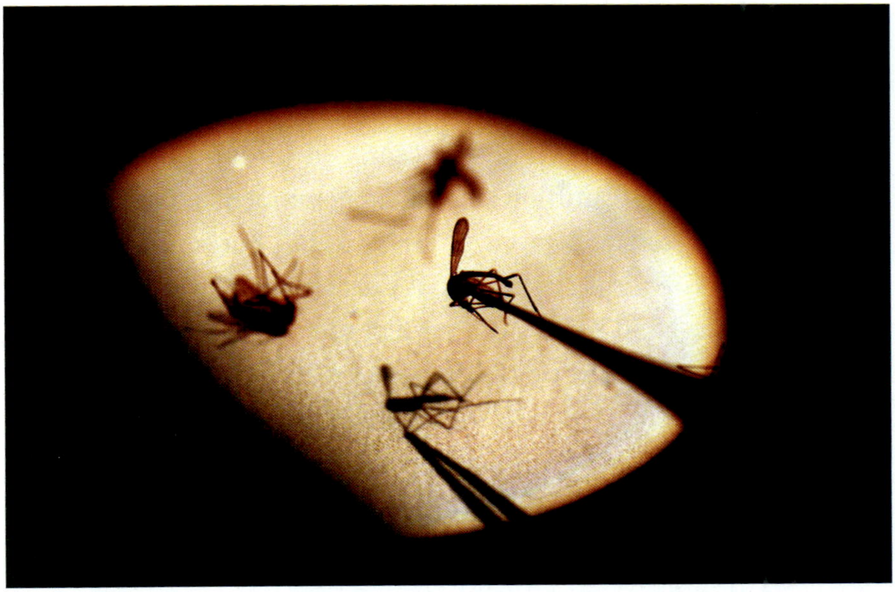

Figure 5.1 Mosquitoes are vectors for many viruses, including West Nile and yellow fever. An entomologist from the American Museum of Natural History discovered mosquito vectors of West Nile virus living in New York City, where West Nile first emerged in the United States.

that the patients at the hospital had SLE and the birds had some other disease. Finally, officials at CDC decided to dig a little deeper. The lab procedure that was used to identify SLE does not just test for SLE. It actually tests for the presence of a whole group of related viruses whose genetic material is made of RNA; SLE is just one member of that group, and was simply seen as the most likely culprit. But there was a chance that the disease was caused by one of the *un*likely viruses. If so, which one?

Scientists at the CDC lab in Fort Collins decided to take a closer look at the virus that the National Veterinary Services Lab had cultured from Dr. McNamara's dead bird tissue samples. Virus hunter Robert Lanciotti used a gene sequencing machine to obtain the sequence of **nucleotides** of part of the mystery virus's RNA. Then he sent this sequence over the Internet to GenBank, a Website that contains a database of genetic sequences (see the box below). Within seconds, GenBank searched through its huge collection of 30 million nucleic acid sequences and found a clear match with a virus never found before in America: West Nile virus (Figure 5.2)! Further study showed that the virus in New York was a perfect match to a variety of West Nile found nearly 6,000 miles (9,600 kilometers) away, in the country of Israel. West Nile virus, which emerged in Africa in the 1930s, had somehow hitchhiked across the

Tools of the Trade: The Internet

The Internet is one of the best tools of modern-day virus hunters. Organizations like the CDC and the WHO have Websites that contain useful general information for scientists studying infectious diseases. Other sites contain information about specific disease-causing microbes. One such site, GenBank, has a huge database of more than 30 million genetic sequences from the nucleic acid of all sorts of organisms, including viruses. The GenBank database was used by one researcher to identify West Nile virus in the 1999 encephalitis epidemic in Queens, New York.

Using the Internet, scientists can easily communicate with each other all over the world. They can e-mail messages on the latest developments in ongoing investigations, and they can discuss ideas in "chat rooms." One such chat room, called ProMED, is used by scientists who study infectious diseases. Researchers used ProMED to compare genetic sequences of different varieties of West Nile virus and determine that the strain that emerged in Queens was identical to the one that caused an epidemic in Israel in 1998.

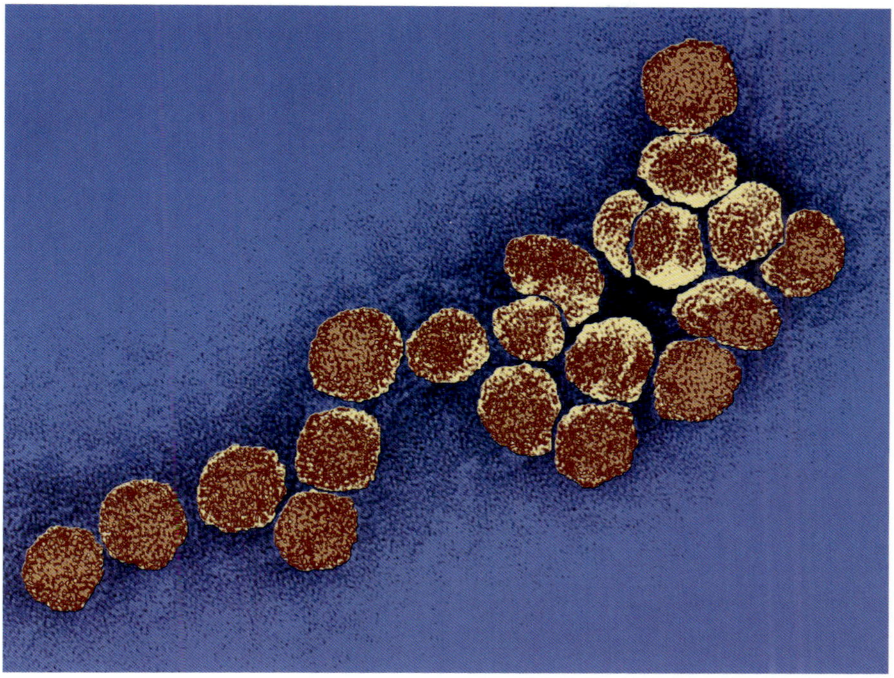

Figure 5.2 West Nile virus particles, seen magnified 125,000 times in this electron micrograph, look very similar to the St. Louis encephalitis virus. Scientists at the CDC at first mistakenly thought that the 1999 West Nile encephalitis outbreak in New York City was St. Louis encephalitis.

Atlantic Ocean. Hidden most likely in a stowaway mosquito or perhaps in an infected bird or person, the virus finally emerged in New York City, where the local mosquitoes spread it to people and birds.

The pieces of the puzzle finally fit together. West Nile virus could infect both birds and people (additional lab tests showed that the patients at Flushing Hospital did indeed have the virus). It could produce all the symptoms seen in the hospital patients, including weak, paralyzed limbs. And it could be spread by the type of mosquito Varuni Kulasekera found living and breeding in Queens.

Possibly the most troubling piece of the puzzle is why it took almost two months to identify the virus.

A LESSON IN COMMUNICATION AND COOPERATION

It was clear to officials at the CDC and elsewhere that many things went wrong during the investigation. Scientists at the CDC initially diagnosed the disease in Queens as St. Louis encephalitis. This diagnosis may have been too strongly influenced by past experience, especially the previous history of SLE outbreaks in that part of the country. As a result, the scientists failed to consider different, possibly unusual, solutions to the problem.

Also, the search to identify the cause of the strange encephalitis in New York City developed into a competition between the CDC and other scientific organizations. This resulted in a lack of communication that slowed down the investigation.

Finally, some people speculate that there may have been a "gender factor" involved. It appears that the concerns voiced by officials at Flushing Hospital and the Bronx Zoo may not have been taken seriously at first because these people were all women, while the CDC scientists in charge of the investigation were all men. This may have resulted in a lack of cooperation that slowed down the investigation.

One would hope that officials at the CDC and elsewhere learned a valuable lesson from this experience: When dealing with emerging viruses, don't let competition and unrelated factors get in the way of clear communication and cooperation.

Chapter 6

SARS: Nightmare in China

IN NOVEMBER 2002, a new disease emerged in the Guangdong Province of southern China. At first, doctors thought they were simply dealing with an ordinary flu outbreak. But it soon became apparent that this was no ordinary flu virus: Along with fever and coughing, victims developed a severe form of pneumonia that made breathing almost impossible. Many patients, even those using machines called **respirators** to help them breathe, died from lack of oxygen, as if they had been suffocated. To make matters worse, the disease was extremely contagious, and many doctors and nurses also became sick.

Perhaps worst of all, Chinese officials, afraid that a public announcement about the strange new disease would cripple Chinese business and tourism, and might cause public panic, tried to keep the outbreak a secret. When officials from the WHO heard rumors about the outbreak and asked the Chinese government for more information, Chinese officials claimed only that they were dealing with a new type of flu. In reality, the outbreak was quickly turning into an epidemic that would soon turn into a global problem.

FROM GUANGDONG TO HONG KONG . . . AND BEYOND

It was inevitable that this highly contagious disease, which we now call SARS (short for "severe acute respiratory syndrome"), would spread to other countries. No one can pinpoint precisely where SARS first emerged, but epidemiologists have been able determine the path SARS took once it spread beyond Guangdong Province.

SARS first emerged outside Guangdong in February 2003, in nearby Hong Kong. The index case was a Chinese physician who had traveled to Hong Kong after treating SARS patients in Guangzhou, a city in Guangdong. The doctor was feeling ill when he checked into Hong Kong's Metropole Hotel, and went to a local hospital the next day. He died from SARS ten days later. Unfortunately, during his brief stay at the hotel, he infected several other guests. Not realizing that they were infected, they left Hong Kong for other countries. The pandemic had begun.

SARS ON THE LOOSE

Three of the SARS-infected guests at the Metropole Hotel returned to their home country of Singapore, where they soon became ill and had to be hospitalized. Because China was still hiding the truth about SARS from the rest of the world, hospital workers did not realize how dangerous and contagious the "flu" they were dealing with really was. Many doctors and nurses quickly became infected, and they, in turn, infected many other people. Although all three of the travelers from Hong Kong eventually recovered, SARS spread to more than 200 people, killing 32, before the true nature of the disease was discovered and Singaporean officials were able to stop its spread.

A similar situation occurred in Canada. Two of the infected Metropole guests were Canadians who brought SARS home with them when they returned to Toronto. They became sick shortly after their return, were hospitalized, and spread SARS to hospital workers. By the time Canadian health officials brought the disease under control, more than 250 people became ill, and nearly 40 died, including the two travelers.

Another Metropole guest was a Chinese-American businessman who became sick on his trip to Hanoi, Vietnam. Shortly after his arrival in Hanoi, he was admitted to the Vietnam French Hospital. When people taking care of the man started becoming ill a few days later, hospital officials called the Hanoi office of the WHO, asking for help. The WHO sent an expert epidemiologist to the hospital to investigate.

CARLO URBANI

Epidemiologist Carlo Urbani was a member of the international medical organization *Medecins Sans Frontieres* (Doctors Without Borders). Urbani's years of studying all sorts of unusual diseases had taught him to expect the unexpected. This knack for diagnosing diseases would quickly prove to be important in snuffing out the SARS outbreak in Vietnam.

When Urbani arrived at the Vietnam French Hospital on February 28, 2003, he immediately realized that the sick businessman and hospital workers did not just have some new kind of flu. This was something much worse. Urbani convinced the Vietnamese government to put the hospital under **quarantine**, to try to stop the disease from spreading. Everyone who was now in the hospital had to stay there, and no one was allowed to enter or leave the building. In addition, special protective suits were provided for hospital staff members who worked with the sick patients.

During his examination of the ill businessman and hospital staff, Urbani collected blood and tissue samples, which were sent to medical laboratories for analysis. It was from these samples that the SARS virus was finally found and identified as the source of the mysterious disease. On March 12, armed with this information, the WHO issued a worldwide alert about this new killer disease.

Despite the precautions taken at French Vietnam Hospital to stop the spread of SARS, 62 people eventually became ill, including Urbani himself. Five of these people died: the traveling businessman, Urbani, and three members of the hospital staff. Urbani died on March 29, barely one month after the Hanoi outbreak began (Figure 6.1). By the time the pandemic ended that summer, SARS had spread to nearly 30 countries, which was easy with today's fast air travel. The disease infected approximately 8,500 people and killed more than 800 (a death rate of just under 10%).

VIRUS HUNTERS IN PURSUIT OF SARS

Since the SARS outbreak was contained in 2003, scientists have continued to study this disease, hoping to develop a vaccine and a cure. This is a tricky problem. Until

Figure 6.1 Dr. Carlo Urbani became infected with SARS while treating SARS patients at Vietnam French Hospital. He died on March 29, 2003. Here, a visitor pays his respects at a United Nations memorial service held in Hanoi, Vietnam's capital city.

researchers learn more about the virus itself, how it infects people, and how the body's immune system tries to fight it off, they won't know the best way to make a vaccine. Other researchers are at work developing treatments and a possible cure for the disease. Some medicines, like the antiviral drug ribavirin, show promise. Perhaps researchers will develop an effective SARS "cocktail cure" by mixing two or more medicines together.

While many scientists are working on developing SARS vaccines and cures, others are trying to determine the source of the disease. Most scientists believe that the virus was "hiding out" in some other animal, or its wastes, that came in close contact with humans. To test this idea, researchers used gene sequencing machines to compare the genes of the SARS virus with viruses found in various wild animals sold as food at Chinese markets in Guangdong. They got a surprising result: A nearly perfect match for SARS genes was found in a virus infecting the civet, a relative of the mongoose! With this information in hand, the Chinese government is expected to stop the sale of civets at markets, just in case they are, in fact, the source of SARS.

Tools of the Trade: Polymerase Chain Reaction

Many virus hunters study the DNA of viruses during the course of their investigations. However, they often need to use more viral DNA than the tiny amount they can obtain from, say, a blood sample. Fortunately, a technique called polymerase chain reaction, or PCR, can quickly produce billions of copies of viral DNA from such a sample in just a few hours.

PCR is a relatively simple technique. The DNA to be copied is added to a test tube containing a solution of nucleotides (the building blocks of DNA), polymerase (an enzyme that assembles DNA strands), and other ingredients. Within just a few minutes, the polymerase is able to double the amount of DNA in the test tube. If there is just one DNA molecule in the original solution, the first doubling, called a "cycle," will produce two DNA molecules. The second cycle will produce four DNAs; the next cycle will produce eight; and so on. Since the cycle can be repeated 30 or more times, a scientist can quickly obtain a large enough amount of DNA to use in his or her research.

Chapter 7

Looking to the Future

RIFT VALLEY FEVER IS a nasty viral disease that first emerged in Africa. In mild cases, a person has little more than a fever for a few days. But severe cases can produce hemorrhagic fever, encephalitis, and blindness. The disease is more dangerous to humans than West Nile encephalitis, and it can be especially virulent when it strikes other animals, particularly livestock such as cattle, sheep, and goats. In 1950, a major epidemic in sheep herds in Kenya killed 100,000 animals. The Rift Valley fever virus (Figure 7.1) is definitely not welcome in the United States!

IT WILL HAPPEN AGAIN

Virus hunters are keeping an eye on the Rift Valley fever virus. Ever since West Nile made its unexpected trip across the Atlantic Ocean to the United States, scientists have feared that the Rift Valley fever virus will do the same, since the epidemiology of these two viruses is so similar: Both first emerged in Africa in the 1930s and eventually spread to countries in the Middle East; both can infect humans as well as other animals; and both are spread by a mosquito vector. As Mike Turell, a scientist at Fort Det-

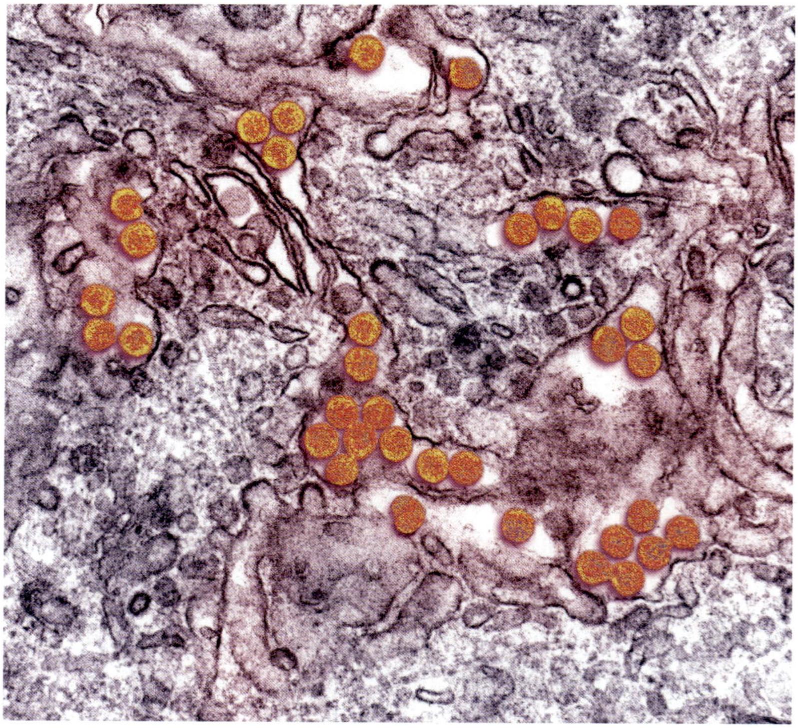

Figure 7.1 This electron micrograph shows deadly Rift Valley fever virus particles in a liver cell of an infected rat. Epidemiologists fear that the Rift Valley fever virus may eventually cross the Atlantic from the Old World to the United States, as the West Nile virus did in 1999.

rick's USAMRIID, describes the situation, "While the chance of any particular mosquito-borne virus hitching a ride on an airplane and surviving in the U.S. is like winning the lottery, somebody always wins the lottery."

MORE AND MORE EMERGING VIRUSES

With all the advances in science, technology, and medicine in the past century, we should be gaining the upper hand in the war against viral diseases. Vaccines have been developed for many viral infections. Smallpox has been wiped out, polio is nearly eliminated, and other viral diseases may one day also be conquered.

In addition, new treatments for viral diseases are continually being developed. In the fall of 2004, for example, a teenage girl in Wisconsin developed rabies after being bitten by a rabid bat. In an experimental procedure, doctors treated their very sick patient with ribavirin and drugs that caused the girl to enter a deep, comatose sleep. A week later, the girl woke up and began a slow recovery. Because the virus caused severe damage to the girl's brain, doctors don't yet know if she will recover completely. But the fact that she survived at all is remarkable: She is the first unvaccinated person *ever* to survive rabies!

Yet, in spite of these impressive accomplishments, the facts speak for themselves: Dangerous viruses keep popping up, sickening and killing people.

Consider Ebola. This deadly virus was unknown before 1976, but since those first outbreaks in Sudan and Zaire, 15 more have erupted in 7 African nations, most recently in Sudan in 2004. Why are all these new viruses emerging, and why do they keep resurfacing? The answer to these questions can be summed up in one word: change.

THE SEEDS OF CHANGE

History has shown that when the "balance of nature" is disturbed, diseases tend to crop up. What happened when people living in tropical grasslands of northern Bolivia cleared the land of natural vegetation in order to build homes and plant crops? The crops provided a huge banquet of seeds and other foods that attracted mice to the area. This resulted in a mouse population explosion in the human settlement. Unfortunately, these rodents were vectors for the Machupo virus. The result: an outbreak of deadly hemorrhagic fever. Similar clearing of rain forests and grasslands by the growing human population in tropical Africa brings people in close contact with hundreds of kinds of animals, one of which is most likely the source of Ebola. In 1971, construction of the Aswan Dam on the Nile River in Egypt flooded a huge area of land behind the dam, creating ideal breeding conditions for mosquito vectors that carry the Rift Valley fever virus. The result was a Rift Valley fever epidemic that made 200,000 people sick.

The way people travel has also changed. The day of the sailing ship is long gone. Modern-day travelers fly all over the world in jet airliners. The SARS pandemic proves how quickly a hot virus can spread: People can become infected in one country, fly halfway around the world to another country, and start an outbreak there before they even know they're sick.

On a larger scale, natural disasters can set the stage for deadly epidemics that sicken and kill thousands. On December 26, 2004, a massive earthquake on the sea floor off the west coast of Indonesia produced a huge ocean wave, or **tsunami**, that crashed onto the beaches of the continents surrounding the Indian Ocean. As of this writing, the death toll was well over 100,000. Scientists fear, however, that more people could end up dying from infectious disease epidemics, arising from decomposing bodies, contaminated water supplies, and unsanitary conditions in crowded refugee camps, than from the wave itself. In fact, within days of the tsunami, outbreaks of measles, a highly contagious viral disease, were already being reported. There is also serious concern that all the standing water left behind by the massive wave will result in epidemics of mosquito-borne viral diseases like West Nile encephalitis and dengue fever, a type of hemorrhagic fever that can kill up to 20% of its victims. The final death toll from this calamity could be unbelievable.

As horrible as the tsunami disaster has become, larger disasters may be just around the corner, caused by the biggest change of all: global warming. Tropical regions of the world seem to be hotbeds of viral infections. HIV, Ebola, West Nile, Marburg, Machupo, yellow fever, and many other viruses first emerged in the tropics. Those that require an animal vector to infect people cannot spread beyond the area inhabited by the vector. The vectors of many viral diseases are insects or other animals that live only in warm, tropical **habitats**. Therefore, diseases carried by these vectors cannot spread into the cooler, temperate regions north or south of the equator. Epidemiologists fear that if global warming continues on its present course, warm, tropical habitats may move farther from the equator, allowing these vectors to expand their range and spread disease to places that used to be "off limits" because they were too cool.

As if all of this isn't enough to worry about, there's more. All the viruses discussed so far in this book occur naturally in the environment. The disease that any virus causes is an incidental *result* of the infection, not the *purpose* of it. The purpose of infection is to allow the virus to survive and reproduce. Imagine if the opposite were true: What if a virus existed solely for the purpose of causing a highly infectious and virulent disease? This is a nightmare virologists hope they never have to face, but they realize that someday, they probably will. Welcome to the world of **bioterrorism**.

THE THREAT FROM BIOTERRORISM

As mentioned in Chapter 2, smallpox was finally eradicated by the WHO in 1977. To be more precise, the *disease* was eradicated; the virus itself was not. It still exists, frozen in special laboratories in the United States at the CDC in Atlanta, Georgia, and in Russia at the State Research Center of Virology and Biotechnology, located in Siberia.

The Russian and American governments have decided not to destroy their stockpiles of the virus. As a result, there is a chance that political and religious terrorist groups have managed to obtain their own secret stashes of smallpox, to be used in terrorist attacks against other countries. Furthermore, these groups may have recruited virologists to create supervirulent varieties of smallpox and other deadly viruses, some of which might be used to make **hybrids**—combinations of, say, smallpox and Ebola. Just try to imagine the consequence of the deliberate release in a large, crowded city, of a deadly smallpox-Ebola hybrid that produced symptoms of both diseases, was as infectious as smallpox, and had Ebola's 90% death rate. The results would be devastating.

Smallpox, Ebola, Marburg, SARS. . . . There are lots of killer viruses out there, available for terrorists to obtain and develop into diabolical weapons of destruction. To try to stay one step ahead of the terrorists, virus hunters hope to be able to improve worldwide **surveillance** against viruses, both natural ones and the modified "designer viruses," in order to detect and stop outbreaks before they can grow into disastrous epidemics and pandemics. To do so will require

Tools of the Trade: Virus Detectors

Several electronic devices currently being developed will someday allow virus hunters to quickly detect and identify killer viruses released by bioterrorists (Figure 7.2). One such device is called an electronic nose; it can actually "sniff" out pathogens in the air. The device consists of tiny fingers or pegs made of a special material that swells up when it detects a specific pathogen, such as smallpox, in the air. The act of swelling breaks an electric current running through the peg, which turns on an alarm.

Another device uses a substance known as a liquid crystal. Microscopic crystals contained in this substance normally line up in a very orderly arrangement, allowing a light beam to pass through the liquid crystal in a straight line. If antibodies for Marburg virus, for example, are inserted into the liquid crystal, they are too small to disrupt the orderly arrangement of crystals. But if Marburg virus particles come into contact with the liquid crystal, the antibodies grab onto the viruses, which are large enough to become wedged in between the crystals, destroying the orderly arrangement. Light passing through such a contaminated, disorderly liquid crystal is now deflected instead of passing straight through. A sensor detects these deflected light beams and turns on an alarm.

Devices such as these, some no larger than a credit card, will someday be placed in crowded places like shopping malls, ready to sound an alarm if terrorists release dangerous microbes. By quickly evacuating people in such an attack, officials may be able to defuse an outbreak before it has a chance to get started.

the development of new, high-tech virus detectors that can quickly identify the viruses that cause these outbreaks (see box on page 71). Placed in the hands of highly trained scientists and technicians, they will give us a fighting chance in the war against bioterrorism.

CONCLUSION

Ever since Edward Jenner developed the smallpox vaccine in the 1700s, virus hunters have been at work. They have been chasing down and studying these microbes, developing vaccines and cures for the diseases they cause. The epidemiologist tromping through tropical rain forests in Africa, searching for the source of Ebola; the virologist working in a modern high-tech virology lab to develop a vaccine against HIV; or the physician tending to a hospital patient

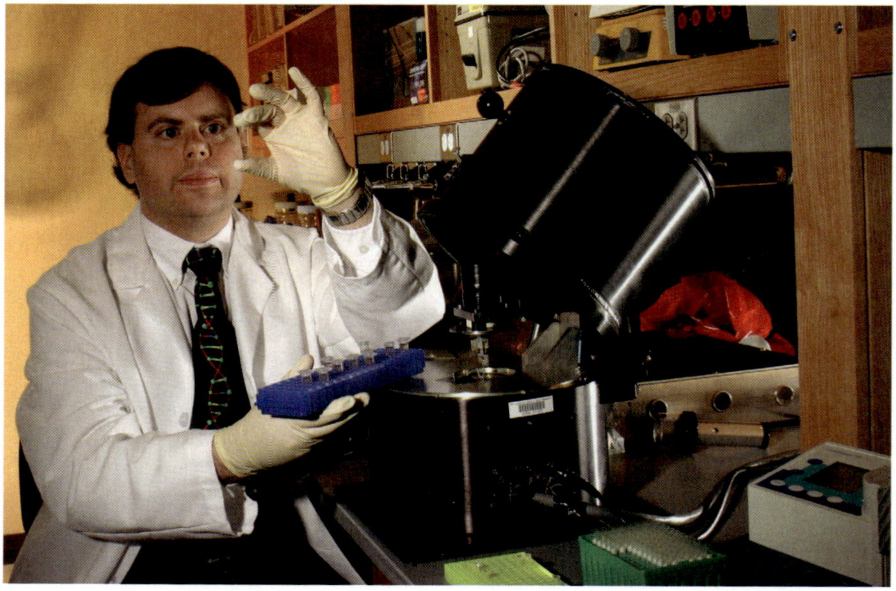

Figure 7.2 This scientist is working with a virus detector, one of many new devices designed to help prevent disaster in the event of a bioterrorist attack.

being given a new experimental treatment for rabies—all of these scientists have the same goal: to reduce the suffering and death viruses bring upon those they infect.

Strange as it may sound, the job of the virus hunter is more challenging than ever before. New virulent viruses are emerging more and more often, and bioterrorists are working hard to produce super-deadly viruses, the likes of which the world has never seen. But the men and women who have devoted their lives to the difficult job of studying these simplest of microbes and conquering the diseases they cause would certainly agree that virus hunting is an exciting, rewarding career. There is no doubt about it: These scientists are up to the challenge!

GLOSSARY

Anthrax: A deadly bacterial disease that can infect many kinds of animals, including humans.

Antibodies: Proteins produced by the immune system when a pathogen is detected. The antibody attaches to the pathogen, labeling it as a target for the immune system to destroy.

Attenuated: Weakened.

Autopsy: The examination and dissection of a human body after death.

Bacteria (singular is *bacterium*)**:** Single-celled organisms that have no nucleus; some bacteria cause diseases, such as anthrax and pneumonia.

Bioterrorism: The use of biological weapons, such as infectious microbes, to attack and harm others.

Botanist: A scientist who studies plants.

Cancer: The growth of abnormal cells in the body that can spread to and destroy healthy organs and tissues.

Cell membrane: A thin sheet-like layer, made mostly of protein and fat molecules, that surrounds a cell.

Comatose: Being in a state of deep sleep or unconsciousness.

Contagious: Easily spread from one person to another.

Convulsion: Violent contracting and relaxing of muscles.

Cytoplasm: The living substance of a cell within the cell membrane, excluding the nucleus.

Diagnosis: The determination of what disease a person (or animal) has by studying the symptoms.

Dispensary: A place where medicine and health care are provided.

Electron microscope: A microscope that uses an electron beam to obtain a magnified image of an object.

Emerge: To come into view or become known.

Encephalitis: Inflammation of the brain.

Endemic: Normally found in a particular area.

Entomologist: A scientist who studies insects.

Enzyme: A type of protein that speeds up chemical reactions in cells.

Epidemic: A widespread outbreak of an infectious disease in which many people become sick at the same time.

Epidemiology: The science of the causes and control of infectious diseases. A scientist who studies epidemiology is called an *epidemiologist.*

Euthanize: To kill an animal suffering from an incurable disease.

Habitat: The place where an organism normally lives.

Hemorrhaging: Losing a lot of blood.

Hybrid: The offspring produced by the mating of two different types of organisms.

Hypotheses (singular is *hypothesis*)**:** Educated guesses or predictions, based on careful observation.

Immune: Protected from disease.

Immune system: A combination of several types of cells and tissues that defend the body against disease.

Immunization: To make immune, usually through vaccination.

Inactivated: Not capable of action.

Index case: The first person to become ill in a disease outbreak.

Infectious: Able to cause disease.

Inoculate: To inject a disease-causing pathogen into the body.

Jaundice: A yellow discoloration of the skin caused by an accumulation of bile, a yellowish waste product, in the blood.

Lesion: A change in an organ or tissue as a result of injury or disease, such as a rash or sore on the skin.

Microbe: A microscopic organism, such as a virus or bacterium.

Mitochondria (singular is *mitochondrion*)**:** Cell organelles that produce energy for cellular processes.

Mutation: A change in an organism's genetic material, producing a new inheritable characteristic or feature.

Necropsy: The examination and dissection of an animal body after death.

Nucleic acid: A complex chemical substance that contains a cell's or a virus's genes. There are two types of nucleic acid: deoxyribonucleic acid (DNA) and ribonucleic acid (RNA).

Nucleotide: A molecular building block of the nucleic acids DNA and RNA.

Nucleus: Cell organelle that contains the genetic material DNA.

Opportunistic infection: An infection by a normally harmless microbe that may become life-threatening if the immune system is weak.

Organelle: A tiny structure within a cell that performs a specific function.

Outbreak: The occurrence of several cases of a disease within a short period of time.

Pandemic: An epidemic that occurs in many areas at the same time.

Paralysis: Loss of the ability to move a body part, such as a leg or arm.

Paralyzed: Showing symptoms of paralysis.

Pathogen: A microbe capable of causing disease.

Pathologist: A doctor or scientist who studies diseases.

Pediatrician: A doctor who specializes in the care of babies and children.

Pneumonia: An infectious disease of the lungs.

Protein: A complex molecule made by cells, composed of amino acids.

Quarantine: A situation in which people are isolated from others to prevent the spread of a contagious disease.

Rabid: Having rabies.

Respirator: A machine or device that helps a person breathe.

Retrovirus: An RNA-containing virus that produces a DNA copy of its genetic material, which is then inserted into a cell's own DNA; from there, it directs the production of more virus particles.

Salivation: The production of saliva in the mouth.

Scanning electron microscope (SEM): A type of electron microscope that passes an electron beam over the surface of an object and produces an image that looks three-dimensional.

Serum: The clear, yellowish liquid part of blood that remains after a clot forms.

Surveillance: A close watch kept over something or someone.

Symptom: A condition that is produced by a disease and can be used to make a diagnosis.

Tissue culture: The growing of plant or animal tissues or cells in an artificial environment.

Transfusion: The transfer of blood from one person to another.

Transmission electron microscope (TEM): A type of electron microscope that passes an electron beam through a thin section of an object and produces an image on a photographic plate or fluorescent screen.

Tsunami: A large, destructive ocean wave caused by an underwater earthquake.

Vaccine: A substance containing dead or weakened disease-causing viruses or bacteria that is introduced into the body to produce immunity to the disease the microbe causes.

Vector: An animal that spreads disease to humans or other animals.

Virologist: A scientist who studies viruses.

Virulent: Very harmful; deadly.

Virus: A disease-causing particle consisting of nucleic acid surrounded by a protein capsule. Some viruses have a fatty envelope surrounding the protein capsule.

BIBLIOGRAPHY

Books and Articles

Abramovitz, Melissa. *West Nile Virus*. San Diego: Lucent Books, 2004.

Casagrande, Rocco. "Technology Against Terror." *Scientific American* 287 (2002): 83–87.

Chee, Y. C. "Heroes and Heroines of the War on SARS." *Journal of the Singapore Medical Association* 44 (2003): 221–228.

Chung, Leung Ping, and Ooi Eng Eong, eds. *SARS War: Combating the Disease*. Hackensack, NJ: World Scientific, 2003.

Despammier, Dickson. *West Nile Story: A New Virus in the New World*. New York: Apple Tree Productions, LLC, 2001.

Drexler, Madeline. *Secret Agents: The Menace of Emerging Infections*. Washington, D.C.: Joseph Henry Press, 2002.

Ezzell, Carol. "Hope in a Vial: Will There Be an AIDS Vaccine Anytime Soon?" *Scientific American* 286 (2002): 40–45.

Facklam, Howard, and Margery Facklam. *Viruses*. New York: Twenty-first Century Books, 1994.

Hecht, Alan. *Polio*. Philadelphia: Chelsea House Publishers, 2003.

McEnery, Regina. "A Call to Arms Against a Goliath of Disease." *The Plain Dealer*. (November 21, 2004): A8–9.

Mezger, Roger. "Making Signs of Infection Crystal-Clear." *The Plain Dealer*. (November 17, 2004): C1, 3.

Morse, Stephen S. "The Vigilance Defense." *Scientific American* 287 (2002): 88–89.

Oldstone, Michael B. A. *Viruses, Plagues, and History*. New York: Oxford University Press, 1998.

Peters, C. J., and Mark Olshaker. *Virus Hunter: Thirty Years of Battling Hot Viruses Around the World*. New York: Anchor Books, 1997.

Ryan, Frank, M.D. *Virus X: Tracking the New Killer Plagues Out of the Present and Into the Future*. Boston: Little, Brown and Company, 1997.

Selim, Jocelyn. "Virus Code Red." *Discover* 26 (2005): 14.

Stewart, Gail B. *SARS*. San Diego: Lucent Books, 2004.

Vallery-Rodot, René. *The Life of Pasteur*. New York: Dover Publications, Inc., 1960.

Villarreal, Luis P. "Are Viruses Alive?" *Scientific American* 291 (2004): 101–105.

Yancy, Diane. *STDs: What You Don't Know Can Hurt You*. Brookfield, CT: Twenty-first Century Books, 2002.

Zimmerman, Barry E., and David J. Zimmerman. *Killer Germs: Microbes and Diseases That Threaten Humanity*. Chicago: Contemporary Books, 2003.

Websites

Centers for Disease Control and Prevention *www.cdc.gov/ncidod/dvrd/spb/mnpages/dispages/rvf.htm.*

Cohen, Jon. "The Mother of all HIV Challenges." *Science*. June 23, 2000. *www.aidscience.org/Science/Cohen288(5474)2160.html.*

Frederick Hecht, M.D. "PCR (Polymerase Chain Reaction)." *MedicineNet.com*. *www.medicinenet.com/script/main/art.asp?articlekey=235 57&pf=3.*

Medecins Sans Frontieres. "Obituary: Carlo Urbani." *www. msf.org/content/page.cfm?articleid=48F5AC68-3D17-4124-9DAFFD3D76597.*

Medilexicon. "Recovery of a Patient from Clinical Rabies, Wisconsin, 2004." *www.medilexicon.com/medicalnews.php?newsid=18307.*

NewScientist.com. "Exotic Market Animals Likely Source of SARS." *www.newscientist.com/article.ns?id=dn3763&print=true.*

Pew Center on Global Warming. "Hurricanes and Global Warming." *www.pewclimate.org/hurricanes.cfm.*

Reilley, Brigg, M. Van Herp, D. Sermand, and N. Dentico. "SARS and Carlo Urbani." *The New England Journal of Medicine*. *http://content.nejm.org/cgi/content/full/348/20/1951.*

Topping, Robin. "The Wave's Deadly Aftermath." *Newsday.com*. *www.newsday.com/news/nationworld/world/ny-c4099595dec30,0,3178211,print.story.*

University of Nebraska-Lincoln. *www.unl.edu/CMRAcfem/em.htm.*

World Health Organization. *www.who.int*

FURTHER READING

Books

Abramovitz, Melissa. *West Nile Virus*. San Diego: Lucent Books, 2004.

Day, Nancy. *Malaria, West Nile, and Other Mosquito-Borne Diseases*. Berkeley Heights, NJ: Enslow Publishers, Inc., 2001.

Durrett, Deanne. *Jonas Salk*. Detroit: KidHaven Press, 2002.

Emmeluth, Donald. *Influenza*. Philadelphia: Chelsea House Publishers, 2003.

Hecht, Alan. *Polio*. Philadelphia: Chelsea House Publishers, 2003.

Hoff, Brent, and Carter Smith III. *Mapping Epidemics: A Historical Atlas of Disease*. New York: Franklin Watts, 2000.

Ridgeway, Tom. *Smallpox*. New York: Rosen Publishing Group, Inc., 2001.

Robbins, Louise E. *Louis Pasteur and the Hidden World of Microbes*. New York: Oxford, 2001.

Serradell, Joaquima. *SARS*. Philadelphia: Chelsea House Publishers, 2005.

Sfakianos, Jeffrey N. *West Nile Virus*. Philadelphia: Chelsea House Publishers, 2005.

Stewart, Gail B. *SARS*. San Diego: Lucent Books, 2004.

Websites

Useful search terms related to virus hunting include scientists' names, disease or virus names, and terms like "research," "symptoms," "vaccine," "epidemiology," and "vector."

CDC Homepage for Rabies: Just for Kids
www.cdc.gov/ncidod/dvrd/kidsrabies/default.htm

Centers for Disease Control (CDC)
www.cdc.gov

Edward Jenner Museum
www.jennermuseum.com/sv/smallpox.shtml

Heroes and Heroines of the War on SARS
http://www.sma.org.sg/smj/4405/4405e1.pdf

National Institute of Allergy and Infectious Diseases (NIAID)
www.niaid.nih.gov.

National Institutes of Health (NIH)
www.nih.gov.

World Health Organization
http://www.who.int

Organizations

The following organizations provide information about infectious diseases found throughout the world. They can provide up-to-date details about ongoing disease outbreaks and epidemics, symptoms of these diseases, research on epidemiology, and treatments and vaccines for infectious diseases.

Centers for Disease Control and Prevention
1600 Clifton Rd.
Atlanta, GA 30333
(800) 311-3435
www.cdc.gov

National Institute of Allergy and Infectious Disease
Office of Communications and Public Liaison
Building 31, Room 7A-50
31 Center Drive MSC 2520
Bethesda, MD 20892
(301) 496- 5717
www.niaid.nih.gov

National Institutes of Health
9000 Rockville Pike
Bethesda, MD 20892
(301) 496-4000
www.nih.gov

World Health Organization
Regional Office for the Americas
525 23rd St. N.W.
Washington, D.C. 20037
www.who.int

INDEX

acquired immunodeficiency syndrome (AIDS) 42, 43, 44, 45, 47
Aguilar, Donato 8–9, 11
American Museum of Natural History 54, 55
Ames, Iowa 53
anthrax 20
antibodies 49
Atlanta, Georgia 70
attenuated vaccine 22
azidothymidine (AZT) 49

bacteria 20
Bangladesh 17
Bay Village, Ohio 51
Bellevue Medical School 25
Bethesda, Maryland 45
bile 11
biosafety level-4 laboratory 39, 40
bioterrorism 70–72
Bolivian hemorrhagic fever 8
  outbreak in Cochabamba, 7–9, 13–14
  spread by mice 14
  symptoms 11
Bronx Zoo 53, 54, 58
BSL-4 laboratory. *See* biosafety level-4 laboratory.

Canada 61
Centers for Disease Control and Prevention (CDC) 36, 44, 53, 54–55, 56, 58, 70
civet 64
Cochabamba, Bolivia 7, 8, 11, 13–14, 33
cowpox 18–19
Cuba 24–25

Democratic Republic of Congo. *See* Zaire.
dengue fever 69
Doctors Without Borders. *See Medecins Sans Frontieres*.

Ebola virus 34, 40, 41, 67, 68, 69, 70, 72
  appearance 37
  funeral ceremonies and 37, 38
  identification of virus 36
  index case 35–36
  outbreak in Sudan and Zaire 34–35, 67
  spread 34–35, 37–39
  symptoms 34, 39
  in United States 39
  vector 38
electron microscope 9, 12, 16, 23, 36, 37, 43, 53, 57, 66
emerging viruses 67, 68–70
encephalitis 51, 54, 65
Enders, John 29
enzyme-linked immunosorbent assay (ELISA) 49
epidemic 15

Flushing Hospital Medical Center 54, 57, 58
Fort Collins, Colorado 53, 55
Fort Detrick, Maryland 9, 39, 67

Gallo, Robert 45–46, 47
gay-related immune deficiency (GRID) 44
GenBank 56
global warming 69
Guangdong Province, China 59, 60, 64
Guangzhou, China 60
Guay, Laura 48

Haiti 45
Hanoi, Vietnam 61, 62, 63
Harvard University 29
Hazelton Research Products 39
hemorrhagic fever 24, 33–34, 35, 36, 37, 65, 68, 69
hemorrhaging 8
HIV 41, 42, 69, 72
  death rate 41
  drugs to treat 48–50
  emergence 44–45
  hiding in immune system 42
  identified 45–47
  killing T4 cells 42, 44
  mother-to-child transmission 48
  mutations 47
  opportunistic infections 42–43, 44
  spread 43, 48
  symptoms 44
  testing for 49
  treatment 47–50
  vaccine 47
Hong Kong 60, 61
"hot" viruses 33
human immunodeficiency virus. *See* HIV.
human T-cell leukemia virus (HTLV) 45, 46

immune system 22, 41, 42–43, 45, 63
inactivated vaccine 29, 31
index case 13, 35–36, 60
Indonesia 68–69
influenza 8, 59
inoculation 19, 22
Internet 56

jaundice 11, 24
Jenner, Edward 18–19, 20, 72
  immunization against smallpox 19
John, prince of England 16
Johnson, Karl 36
Jones, Cenydd 35–36

Kenneth A. Scott Wildlife Rehabilitation Center 51
Kidenya, Jayo 47–48
Knoll, Max 12
Kulasekera, Varuni 54, 57

Lanciotti, Robert 55–56
Lincoln, Abraham 16
Los Angeles, California 44
Louis XV, king of France 16

Machupo virus 9, 24, 33, 40, 41, 68, 69
Marburg, Germany 36
Marburg virus 36, 37, 40, 69, 70, 71
Marcus Aurelius 16
Maridi, Sudan 34–35, 36
McNamara, Tracey 53, 54, 55
measles 69
Medecins Sans Frontieres 61
Meister, Joseph 22–23
Metropole Hotel 60, 61
Montagnier, Luc 46–47
Murphy, Fred 36
mutations 47

National Institutes of Health (NIH) 45
National Veterinary Services Lab 53, 55
Nelmes, Sarah 19
nevirapine 48
New York, 25, 44, 52, 53, 54, 55, 57, 58
Nzara, Sudan 34, 35, 36

opportunistic infections 42–43, 44
oral vaccine 30

pandemics 40, 45, 60, 68, 71
Paris, France 20
Pasteur Institute 46
Pasteur, Louis 20–23
pathogens 42, 43
Peters, C. J. 9
Philadelphia, Pennsylvania 24
Phipps, James 19
poliomyelitis 15, 28, 67
  cause 28
  infantile paralysis 28
  infection rate 31–32
  symptoms 28
  vaccine development 28, 29–30, 31, 36
polymerase chain reaction 64
Prince of Asturia Prize for Scientific Research and Technology 46
ProMed 56
protease inhibitor 49

quarantine 62

rabies 15, 19–23, 41, 67
  death rate 21, 23
  development of vaccine 21–23
  symptoms 21
Reed, Walter 25–27
Reston, Virginia 39
retrovirus 42, 43, 45, 47
reverse transcriptase (RT) 45–46
ribavirin 67
ribonucleic acid (RNA) 42, 47, 55
Rift Valley fever 65, 66, 68
Ruska, Ernst 12
Ryan, Frank 34

Sabin, Albert 30, 31
St. Louis encephalitis (SLE) 53, 54, 55, 57, 58
Salk, Jonas 29–30, 31
San Francisco, California 44
SARS
  Chinese government secrecy 59–61
  identification 61–62
  source 63
  spread to Canada 61
  spread to Singapore 60–61
  symptoms 59
  treatment 63–64
  2003 outbreak 59–62, 68
  WHO travel warnings 62
scanning electron microscope (SEM) 12

Seton Hospital 7, 8, 9, 13
severe acute respiratory syndrome. *See* SARS.
Singapore 60
smallpox 15, 16, 70
- death rate 18
- eradication 18, 19, 67, 70
- lesions 16–18
- long-term effects 17
- spread of 16–18
- symptoms 16–18
- vaccine 72

Smith, David 35
Smolensk, Russia 19–20
Somalia 18
Spanish-American War 24–25
State Research Center of Virology and Biotechnology (Russia) 70
Sudan 34, 67

Tadiello, Megan 52
T4 cells 42, 43, 44, 45
Theiler, Max 27
tissue culture 29
transmission electron microscope (TEM) 12
tsunami (December 26, 2004) 68–69
Turrell, Mike 66–67

UCLA Medical Center 44
Uganda 47, 48
United Nations 63
University of Cincinnati 30
University of Pittsburgh 29
Urbani, Carlo 61, 62, 63
U.S. Army Medical Research Institute of Infectious Diseases (USAMRIID) 9, 67

vaccines 12–13
vectors 25, 26–27, 55
Vietnam French Hospital 61, 62, 63
virus 8, 9–11
- epidemiology 13–14
- reproduction 11
- structure of 9–10
- symptoms of viral disease 11
- virus identification 11, 12–13

virus detector 71–72

Walter Reed Army Medical Center 27
Washington, D.C. 27, 39
Washington, George 16
Webb, Patricia 36
West Nile virus 51, 65, 66, 69
- identification 52, 55–57
- 1999 outbreak 52, 54–58
- problems with outbreak investigation 58
- spread by mosquitoes 54, 55, 57
- symptoms 51–52, 54, 56, 57

World Health Organization (WHO) 18, 19, 24, 56, 60, 61, 62, 70

Yambuku, Zaire 35
yellow fever 15, 23–27, 69
- cause 26
- death rate 24
- prevention 27
- spread by mosquito 24, 25–27, 55
- vaccine 27

Yellow Fever Commission 25–27
Yusia 34, 35–36

Zaire 34, 35, 67

PICTURE CREDITS

page:

10: © Peter Lamb
17: © CDC/PHIL/CORBIS
23: © Dr. F. A. Murphy/Visuals Unlimited
27: © Bettmann/CORBIS
31: © Bettmann/CORBIS
37: © Science VU/CDC/Visuals Unlimited
38: © GILBERT LIZ/CORBIS SYGMA
43: © Dr. Hans Gelderblom/Visuals Unlimited
46: © Reuters/CORBIS
55: © WYMAN IRA/CORBIS SYGMA
57: © Science VU/CDC/Visuals Unlimited
63: © Reuters/CORBIS
66: © T. W. Geisbert/Visuals Unlimited
72: © Rick Friedman/CORBIS

Cover: © Liu Dawei/Xinhua Photo/CORBIS

ABOUT THE AUTHOR

RICK EMMER, formerly a professional aquarist at the Cleveland Aquarium and a zookeeper at the Cleveland Metroparks Zoo, is now a substitute science and math teacher for the Avon Lake City School District in Ohio. He holds two degrees in biology: a Bachelor of Science from Mount Union College and a Master of Science from John Carroll University. He lives in Bay Village, Ohio, with his wife, Liz; three children; two guinea pigs; a green anole lizard; and a tree snail—not to mention countless invisible virus particles. Emmer has written numerous magazine articles about natural history and animal husbandry. *Virus Hunter* is his first book.